D0853826

No. 1006
$7.95

BUILD-IT BOOK OF SOLAR HEATING PROJECTS

BY WILLIAM F. FOSTER

FIRST EDITION

FIRST PRINTING—OCTOBER 1977
SECOND PRINTING—JUNE 1978

Printed in the United States
of America

Hardbound Edition: International Standard Book No. 0-8306-9996-1

Paperbound Edition: International Standard Book No. 0-8306-1006-5

Library of Congress Card Number: 77-21198

Cover photo courtesy of Dave Wood and separations from Farnam Equipment Co., Phoenix AZ.

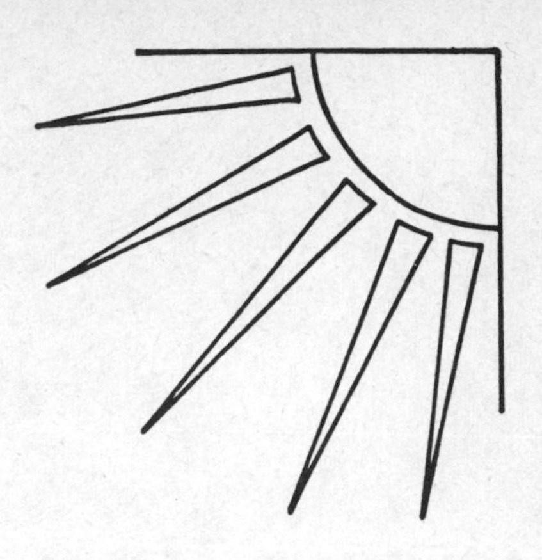

Preface

Continuing inflation, coupled with concerns over the cost and availability of residential heating fuel, has encouraged many innovative homeowners to build their own solar heaters. Results range from attractive, well-engineered projects to unsightly assemblages yielding disappointing results.

This book is designed to give those homeowners who have the requisite skill, the know-how to construct and install systems properly. The designs suggested in Chapters 6 through 9 represent typical solutions to heating problems. The performance of the systems will depend upon local climatic conditions and the quality of construction. Compliance with local building codes is essential.

The *Homeowner's Guide to Solar Heating & Cooling,* TAB Book No. 906, is suggested as prerequisite reading. It delves into the economics and theory of solar heating, providing the homeowner with key factors to assist in making the investment decision.

William M. Foster

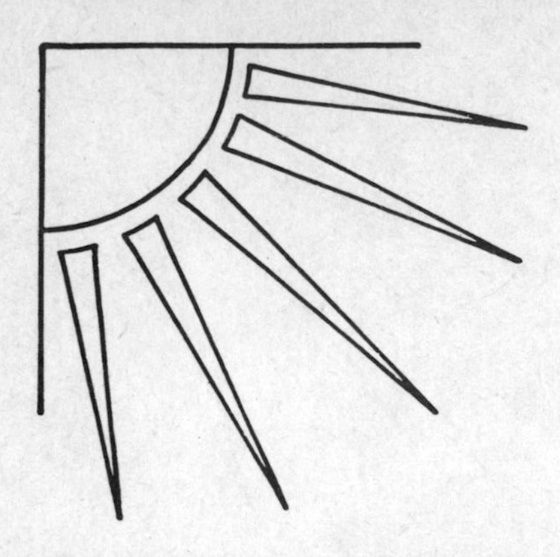

About the Author

William M. Foster received his engineering degree from the University of California at Los Angeles. He is Corporate Manager of Solar Programs for Alcoa and is active in several trade associations as a solar advisor.

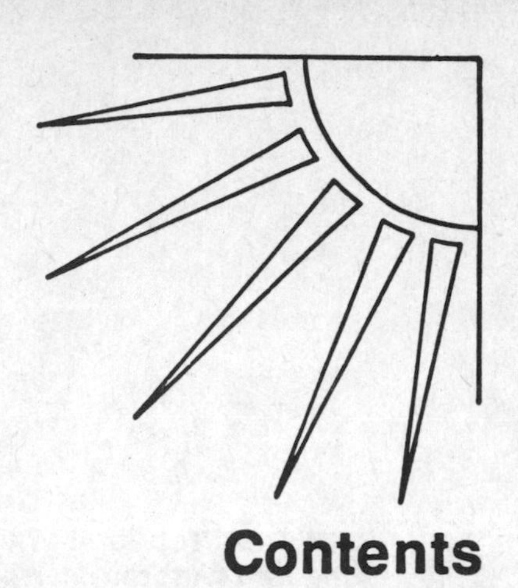

Contents

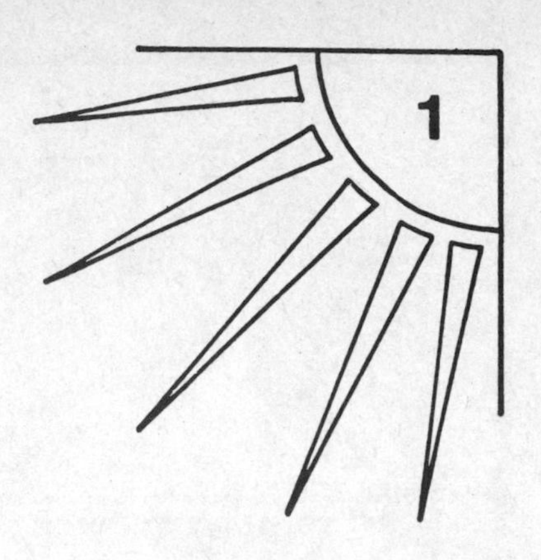

Requisite Skills

In this chapter, I will describe the metalworking, joining, and glazing operations necessary to construct a typical solar collector. In addition, I will outline the basic plumbing and electrical tasks required for system assembly. After reading the chapter, you should have a good idea of whether you have the tools and skill to make your own collectors or the ability to assemble factory-produced units.

Knowledge of how a solar heater operates is necessary before any construction begins. Although this theory is covered thoroughly in my *Homeowner's Guide to Solar Heating & Cooling*, TAB Book No. 906, a brief review follows.

UNDERSTANDING THE PRINCIPLES

There are four main elements in a solar heating system:

- collector
- storage tank
- auxiliary heat supply
- control system

Sunshine heats the *collector*, which is an insulated metal box covered by glass or other transparent material. This covering enables heat to build up inside the collector. Otherwise the heat would be lost to the atmosphere by radiation and by convection caused by wind blowing over the metal surface. The sides and back of the box are insulated to retain the

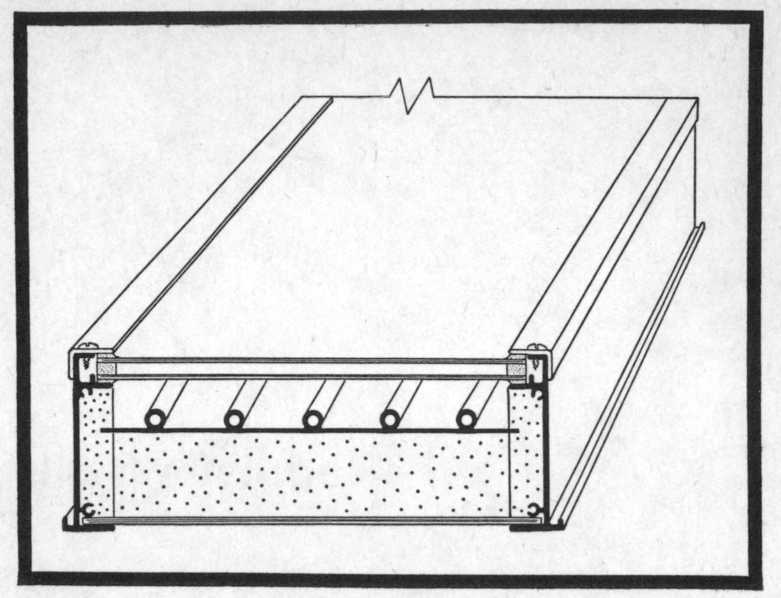

Fig. 1-1. Solar collector assembly. Metal absorber circulates fluid. It is well insulated and covered by two panes of glass.

trapped heat. With nothing to carry this heat away, temperatures inside the collector can reach 400°F in hot climates. This heat is wasted unless it can be removed from the collector on a systematic basis and can be either used directly or stored for future use.

Heat removal is accomplished by circulating fluid through passageways in the metal *absorber* contained inside the collector (Fig. 1-1). As long as the fluid is colder than the absorber, it will collect heat, and its temperature will rise. The degree of heat transfer is controlled by the material and construction of the absorber, the rate of fluid flow through the absorber passageways, and the type of fluid selected.

Storage of the heat is necessary, because demand for hot water or space heating generally does not coincide with the best hours for collecting solar heat, which are typically 9:00 a.m. to 4:00 p.m. *Storage tanks* may be sized to provide solar-heated water over a two-day period with little or no intervening sunshine. Storage for a longer period may be questionable from an economic standpoint.

Auxiliary energy is needed to supplement the solar system whenever the heat being stored falls below the service or use temperature. For a hot water heater this is generally 120°F to

140°F, and it is 80°F to 90°F for space or room heating. When the heat being stored falls below these temperatures, the auxiliary heating system using gas, electricity, or fuel oil boosts the temperature. The object of solar heat, therefore, is not to provide 100 percent of your annual heating needs, but to reduce the conventional heating bill enough to merit the investment.

With the exception of a solar hot water heater operating on the thermosiphon principle (Chapter 6), all the systems in this book require circulating pumps and controls. The heart of the *control system* is a device called a *differential thermostat*. It uses two heat sensors: one measures the temperature of the fluid leaving the solar collectors, and the other measures the storage tank temperature. When the collector outlet temperature exceeds that of the storage tank by a specified amount (generally 20°F), a relay in the unit closes, starting the collector pump motor. The pump will continue to operate, circulating fluid through the collectors, as long as heat can be contributed to storage. There is no pumping action on overcast days or at night, because such action would only lose heat to the atmosphere.

Figure 1-2 and the explanation that follows will describe the typical operating cycle of the thermostat for one day's operation.

The collector pump is initially *off*. As the sun rises at 7:00 a.m., the collector temperature increases rapidly above the

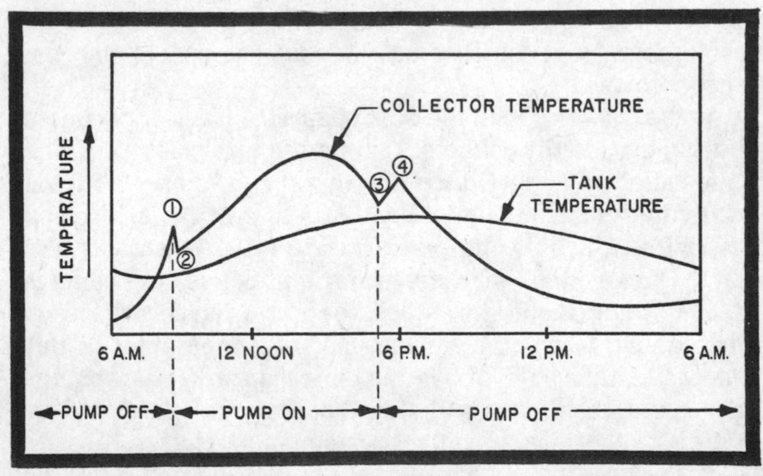

Fig. 1-2. Daily operating cycle of a circulation pump controlled by a differential thermostat.

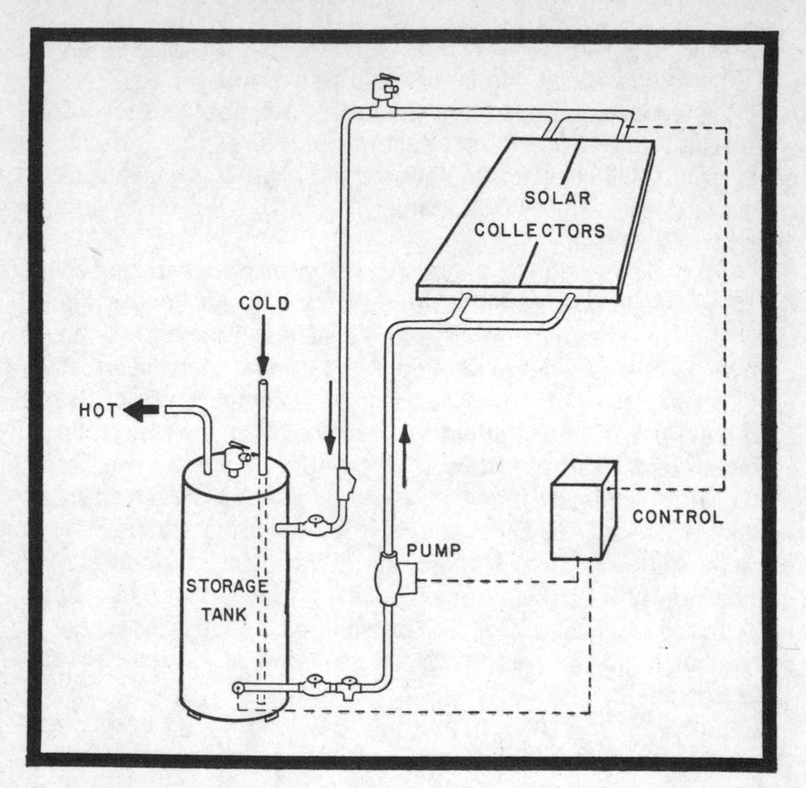

Fig. 1-3. Hot water preheater design using a circulation pump. No freeze protection is provided.

storage tank temperature. At point 1 (Fig. 1-2), 20° F above the tank temperature, the thermostat relay closes, starting the pump motor.

As the cooler fluid in the tank begins to circulate, it causes an initial drop in the collector temperature, to point 2. As long as the collector temperature remains 3° F or more above the tank temperature, the pump continues to operate. As evening approaches, the collector temperature falls. When it is less than 3° F above the tank temperature, at point 3, the pump is turned off. Once fluid flow stops, there is a temporary rise in temperature, to point 4. As long as the temperature at this point is less than 20° F above the temperature of the tank, the pump remains off, and will stay that way until sunrise. This differential-thermostat control system has factory settings to assure that maximum heat energy is stored and retained in the tank under various weather conditions.

A hot water preheater using these controls is illustrated in Fig. 1-3. Here the collectors are operating at line pressure and are circulating potable water. Material selection and assembly are most important in a system of this type. This unit is designed to operate in climates where freezing temperatures do not occur.

I mentioned a thermosiphon system earlier; it is the simplest type of solar water heater to understand and build. In this system, the storage tank is located above the solar collectors. If the top of the collector is one to two feet below the bottom of the storage tank and the connecting lines are well insulated, heated water should flow upward into the storage tank and be replaced by colder water with increased density. With proper system design, this flow occurs without the aid of a pump and will not reverse at night when the collectors cool off.

Figure 1-4 describes the flow path. This is the type of system used in most early Florida installations and is

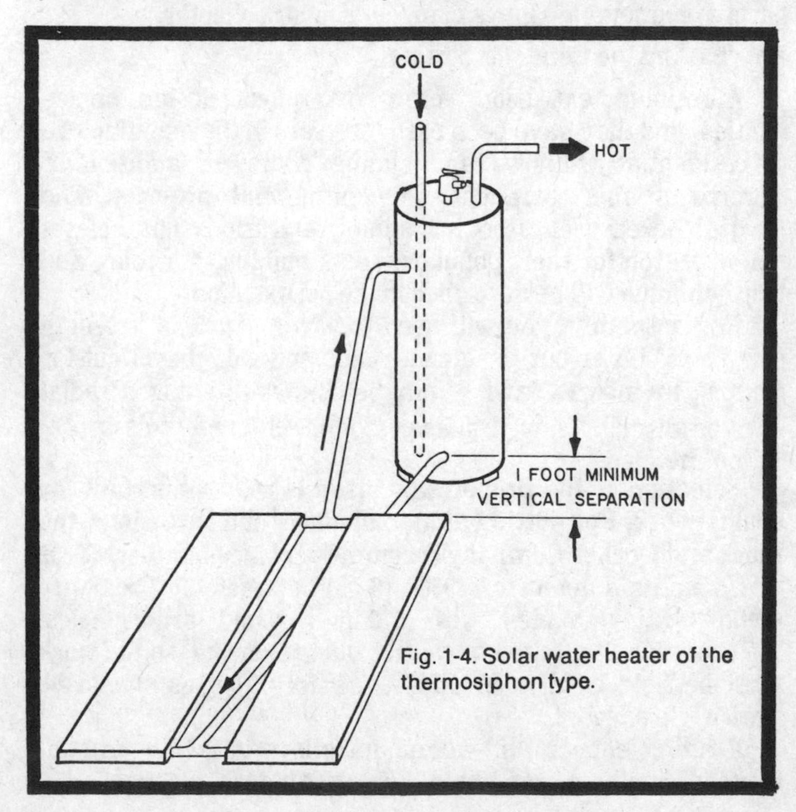

Fig. 1-4. Solar water heater of the thermosiphon type.

currently quite popular in Australia. Cold water is supplied at city main pressure, which can be up to 100 pounds per square inch, so all connections and components must be capable of withstanding that pressure.

METALWORKING

From a durability standpoint, I do not believe in using exposed wood for solar collector frames. The designs in this book employ an extruded aluminum framing system. Wood is used only for the back plate of the collector, and exterior or marine grade plywood is specified for this purpose. I prefer to use aluminum-faced plywood manufactured by Weyerhaeuser, called *Panel 15*.

Absorber plates (when constructed, rather than purchased as an assembly) will use aluminum and copper in both sheet and tubular forms. You will need the tools and techniques for cutting, drilling, bending and, of course, joining these materials. If you refer to Fig. 1-1, you can see how the different parts are assembled into a completed solar collector.

Fabricating the Collector Frame

Aluminum extrusions offer marvelous design opportunities, and they have been used for years in the manufacture of residential windows and sliding doors, in addition to storefronts and entrances for commercial projects. The *integral screw slot* used for joining extrusions has been a major reason for their popularity as a building material, and this technique will be described in the next section.

For assembly, you will need to saw extrusions to length using a 45° miter cut for glazing caps and a 45° bevel cut for framing members. Sawing may be done with either a radial arm or circular saw. Hacksaws or other hand tools are not recommended.

Selection of the proper saw blade is most important for quality work. For cutting aluminum sheet and extrusions, the Black and Decker Company recommends blade numbers 39602 (7 1/4-inch diameter) or 39608 (8-inch diameter). These are metal cutting blades with 5/8-inch round arbor holes. Thin-gauge metal tends to chatter during sawing, so the work must be held tightly. In addition, safety goggles should be worn while sawing.

Drilling copper and aluminum alloys is fairly straightforward, using a hand-held 1/4- or 3/8-inch electric drill.

Always use a sharp center punch to mark the point of entry of the drill and to prevent it from wandering. I usually place one drop of cutting oil in the punch mark and use a general-purpose, high-speed steel twist drill with a steady pressure applied while cutting.

Fabricating the Absorber Plate

For the collector designs in this book, cutting, reaming, bending, and drilling are the only metalworking operations required on the absorbers. Small diameter copper and aluminum tube may be cut to length using a hand-held tubing cutter, pictured in Fig. 1-5. (Figures 1-5 through 1-13 are reproduced with the permission of NIBCO Inc.) The cutting wheel is tightened against the tube where the cut is to be made and is continually tightened as the cutter is rotated around the tube perimeter. This gives a good square cut but produces a burr inside the tube which must be removed. Most cutters have a built-in reamer which will do the job. The reamer is rotated inside the tube, as pictured in Fig. 1-6. If you do not have a reamer, a round or 1/2-round file may be used.

Large diameter tube may be cut with a power saw, using the metal cutting blade previously suggested for aluminum, or with a large tubing cutter if you have one. If you use a hacksaw, be sure to make a *square* cut. Both the inside and

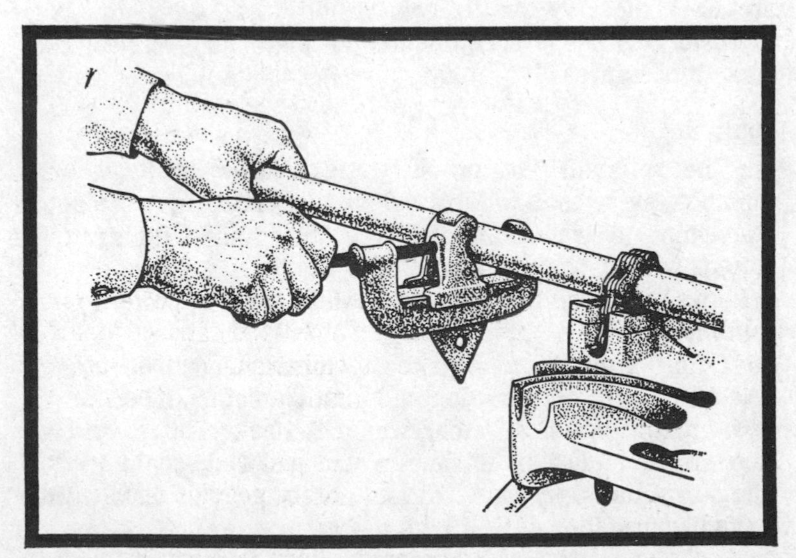

Fig. 1-5. Cutting copper water tube with a roller cutter. Tighten the roller slowly to minimize deflection.

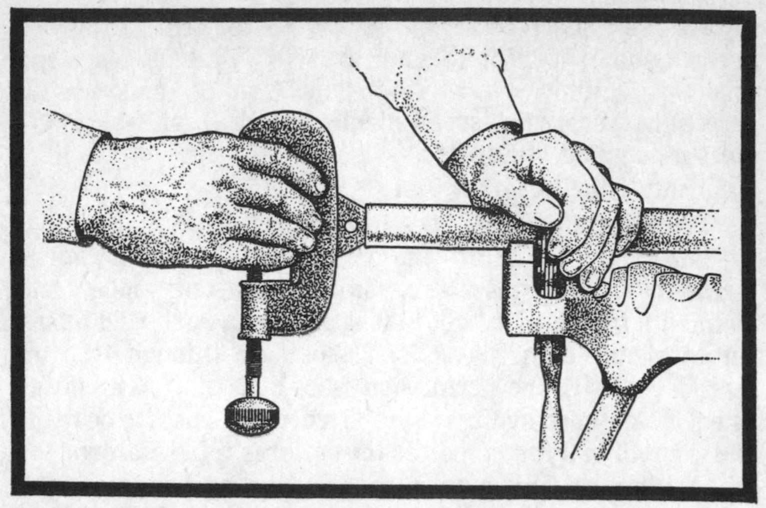

Fig. 1-6. Use of a flat metal reamer to remove burrs from the inside diameter of water tube.

outside diameters of the tube must be filed after cutting to remove burrs.

JOINING METALS

Only two joining techniques will be discussed in this section: soldering and mechanically fastening aluminum extrusions by use of screw slots and self-tapping screws. The most important skill is making pressure-tight solder bonds.

Soldering

The reliability of properly soldered joints in copper plumbing systems is well established. In fact, from a strength standpoint, a well-made solder joint between tubing and fittings will be stronger than the tubing itself. Achieving good results, however, depends upon your skill in producing a *well-made joint*.

Before attempting to make any soldered joints, be sure to select the right composition and quality of *flux* and *solder*. I recommend type 50A solder which is 50 percent tin and 50 percent lead (40A is 40 percent tin and 60 percent lead). Bargain solders generally have 40 percent tin or less; avoid them because they are difficult to flow or draw into the joints, they are pasty, and they generally do not flow as well as the 50/50 types. I suggest using a paste-type solder flux; these are generally made of petroleum jelly containing a mild acid. If you have a choice, pick the same manufacturer for both the flux and the solder, to insure compatibility.

Some of the solders and fluxes sold in hardware stores yield disappointing results. I suggest you try sample joints using a small quantity of two or three different products until you find one you are satisfied with. If you can make joints that show a continuous fillet of solder *inside* the fitting at the end of the tube, you will probably have good results in joining a solar absorber assembly. Soldering aluminum is more difficult, but it can be done with the proper technique.

Here are the techniques for preparing and soldering copper tubular joints. After cutting and deburring, the tube and fitting must be properly cleaned. To remove dirt, oxide, and oil accumulated from handling, clamp the tube in a pipe vise and use a fine grit emery paper to clean the entire perimeter, as in Fig. 1-7.

If you look inside a well-made tube fitting, you will see a ridge that stops the tube when the parts are assembled. You should clean the tube 1/4 to 1/2 inch *more* than the depth of this socket, or 3/4 to 1 inch from the end of a 1/2-inch diameter tube. Again use fine grit emery paper, as in Fig. 1-8. Clean all the way down to, and including, the bottom of the socket where the ledge starts, but don't remove more metal than is necessary to get a clean-looking surface.

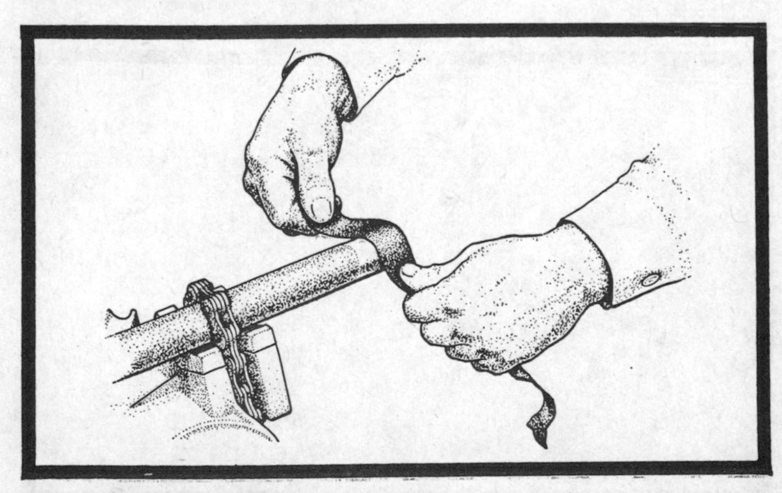

Fig. 1-7. Cleaning copper water tube using emery paper. A fine grit paper is preferred. Clean the entire perimeter.

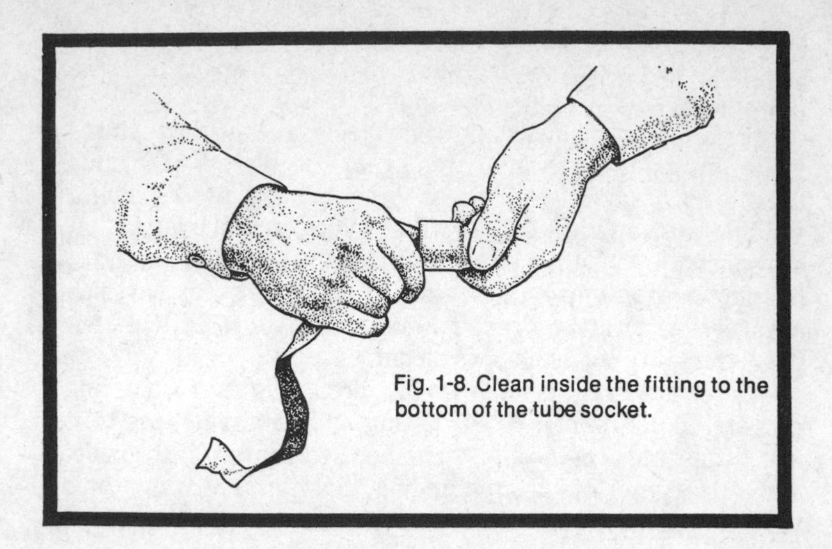

Fig. 1-8. Clean inside the fitting to the bottom of the tube socket.

Next, *flux* must be applied over the cleaned surfaces. Flux protects the cleaned metal from oxidizing when the parts are heated. If they did oxidize, the solder would not stick. The fluxes generally used come in paste form and can be applied with a small, stiff brush made for this purpose. Apply flux in a *thin*, even layer to the tube (Fig. 1-9) and fitting. Be sure to use a clean brush, and cover *only* those areas to be soldered. Fluxing and soldering should be done within an hour or so after cleaning the parts.

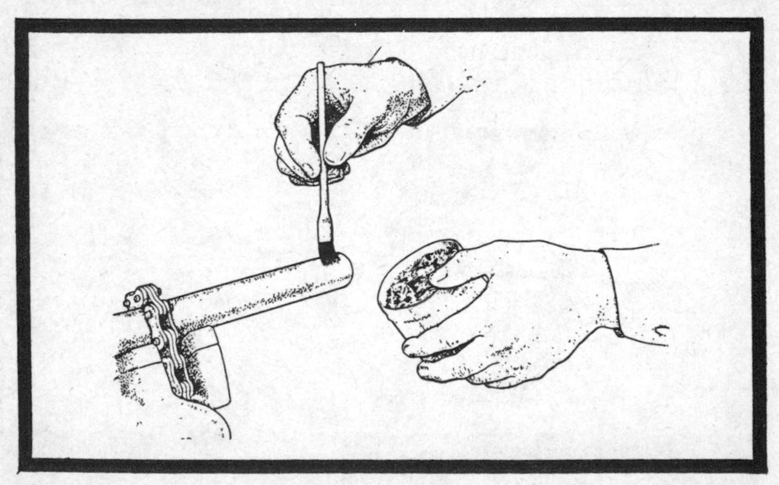

Fig. 1-9. Application of paste flux using a clean stiff brush. Apply sparingly and flux only the area to be soldered.

Assemble the parts, making sure the tube is inserted into the fitting until it *bottoms out*. After assembly and correct alignment, apply heat as described next.

The process of heating is the point at which most people go wrong when soldering; skill is required to do this job properly. Although acetylene or natural gas torches may be used, I prefer a simple propane torch with a flexible rubber extension connected to a standard 14-ounce cylinder. This is inexpensive and available at most hardware stores. I would start by using a torch tip that delivers a relatively small flame size; then as you learn to make good joints, a larger flame and shorter heating periods may be used.

Heat the *tube* first, directing the flame slightly away from the fitting, as illustrated in Fig. 1-10. Try to keep the flame away from the fluxed and cleaned areas to minimize oxidation. After heating the tube momentarily, alternate the flame between the tube and fitting (Fig. 1-11). Heat transfer is best when the tip of the blue *inner* cone of the flame just touches the metal.

You should test the heat level frequently by touching a piece of *solder* to the junction of the fitting and the tube. *Never* melt the solder with the flame. It will melt by contact with the copper when the proper temperature is reached. When the solder begins to melt, move the flame to the corner of the fitting, rotating it as necessary to heat the entire perimeter of the part. Feed the solder into the joint (Fig. 1-12) all the way around the tube. To give you an idea of how much solder it takes to fill the joint, bend the solder roll (see Fig. 1-12 again) after measuring 1 inch. This is all it should take to make one

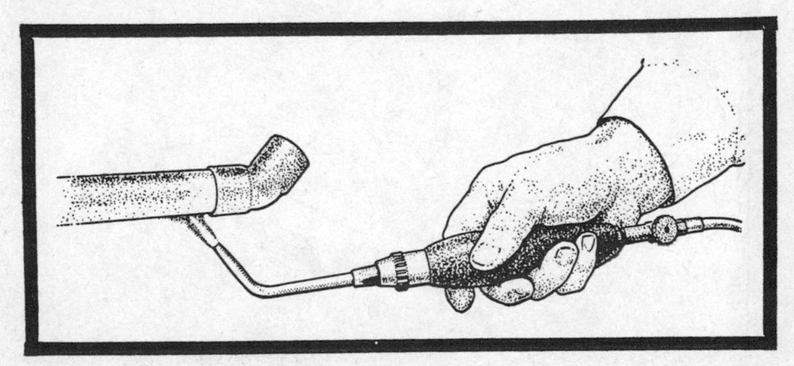

Fig. 1-10. Use a torch to apply heat to the joint. Heat the tube first, keeping the flame away from the fluxed area.

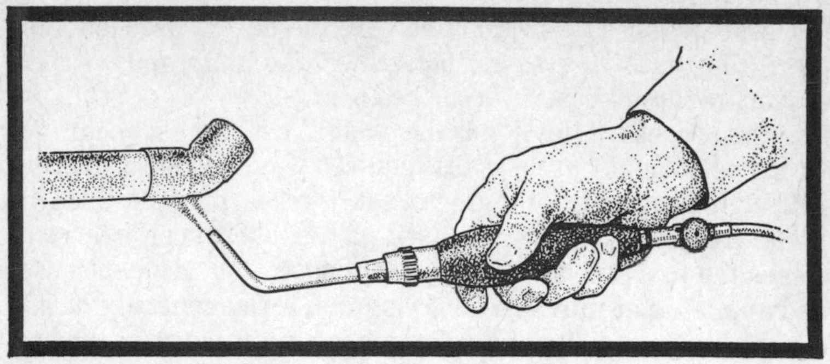

Fig. 1-11. Transfer heat to the fitting, alternating the flame to insure all areas are up to temperature.

joint on a 1/2-inch diameter line using 1/8-inch diameter wire solder. Once the joint is full of solder, remove the flame. Overheating may damage the joint.

Completed joints may be wiped with a clean cloth (Fig. 1-13), while they are still hot, to remove excess solder; but this is not essential. If you do wipe it, be careful not to *move* the parts while the solder is solidifying, or a cracked joint will result.

Mechanical Fastening

The easiest method of joining extruded collector frames is with self-tapping or sheet-metal screws. There are standard

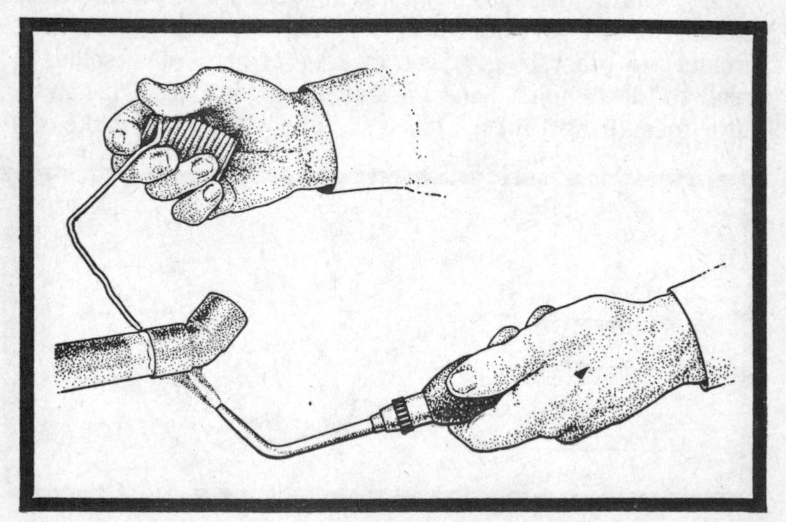

Fig. 1-12. Apply solder to the heated joint. When metal is at the proper temperature, solder will melt upon contact with the joint.

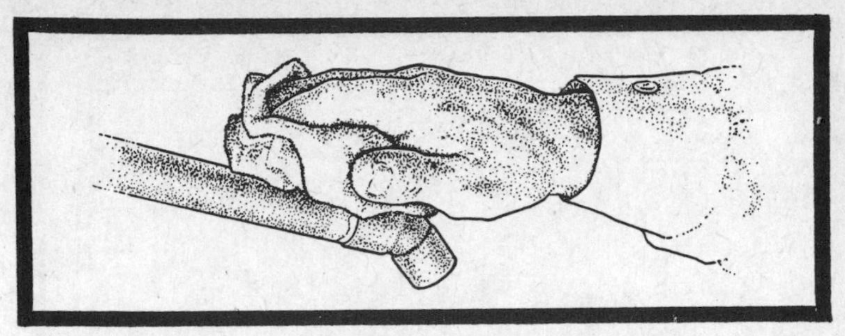
Fig. 1-13. To improve appearance, solder joints may be wiped with a clean cloth while still hot.

frame extrusions available, which you can buy through metal distributors, that incorporate these features in their design. You will note the Alcoa framing system detailed in Fig. 1-14 has two integral screw slots in the frame. If you miter the

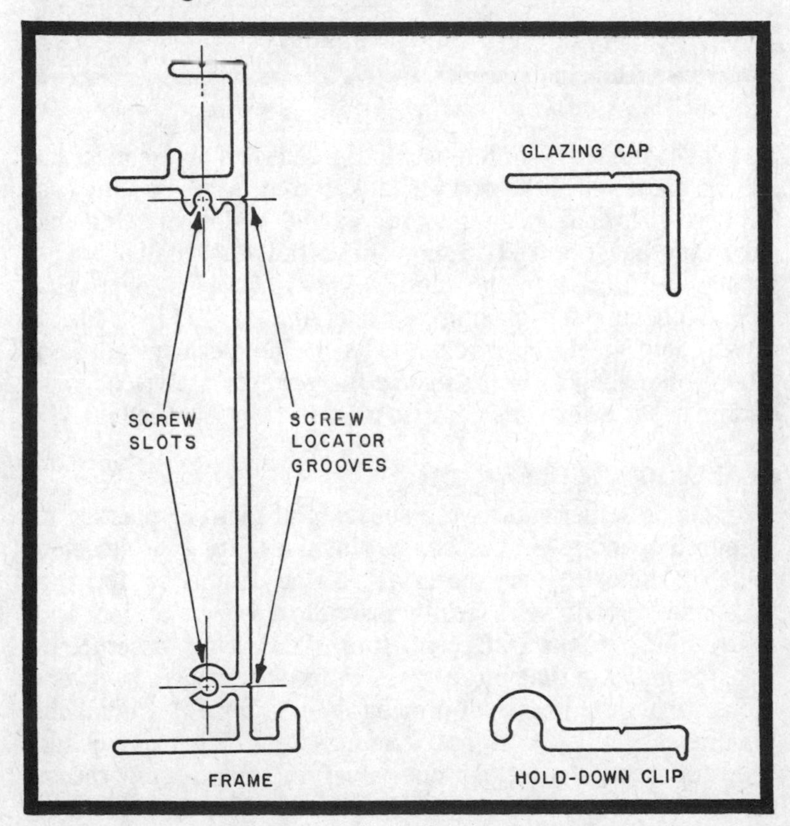

Fig. 1-14. Three extrusions used in the Alcoa framing system. Grooves to locate screw slots are provided.

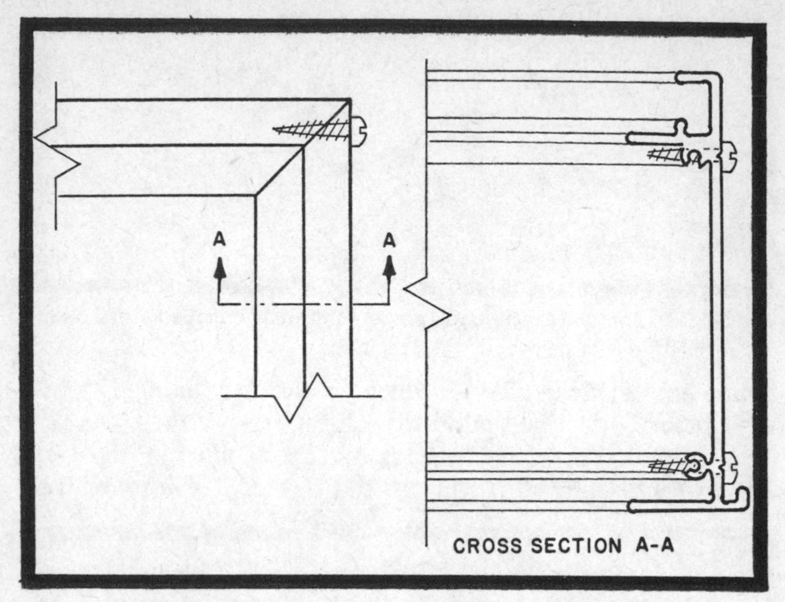

Fig. 1-15. Assembly of frame members using integral screw slots.

parts at 45° and drill a pilot hole in the end sections equal to the outside diameter of the screw shank, the parts will be joined as the screw threads itself into the extruded slot, drawing the parts together. Figure 1-15 show this detail. The small *notch* or groove extruded into the metal (Fig. 1-14) helps locate the drill. To secure the glazing cap over the glass (Fig. 1-16), a self-tapping sheet-metal screw is used. The Alcoa system also offers a hold-down clip to secure the panel to a plywood roof deck in areas where this detail is permitted by local codes.

GLAZING THE COLLECTOR

Glazing is defined as the securing of glass or plastics in prepared openings. In this case, it involves fitting the selected solar collector cover material to the frame to form a weather-resistant seal, while providing for expansion and contraction of the different materials being assembled. Reports indicate that more glass failures on collector covers occur because of improper glazing than because of vandalism. Plastics generally will not fracture, but they expand and contract more than glass does over the same temperature range. Remember that as you tilt glass or other flexible cover materials from a vertical position, their weight causes them to start deflecting or sagging. Therefore, the *size* of the piece is an important consideration. For this reason, I have limited collector designs using glass covers to about 20 square foot in size

Selection of Cover Material

Glass is not the only option for solar collector covers. The thin-film plastics, also called polyvinyl fluorides (PVF), are available in 0.004 inch thickness. Such a material is described as an *economy option* for the system in Chapter 9. fiberglass-reinforced plastic that has been properly stabilized to reduce ultraviolet degradation and provide good weatherability may be suitable for some applications. Cutting and handling of this material is relatively simple, and using the material provides flexibility by enabling you to increase the size of the collector.

In the final analysis, glass is the most popular choice for solar collector covers, however. It is easy to clean, noncombustible, relatively strong, and has good thermal and ultraviolet stability.

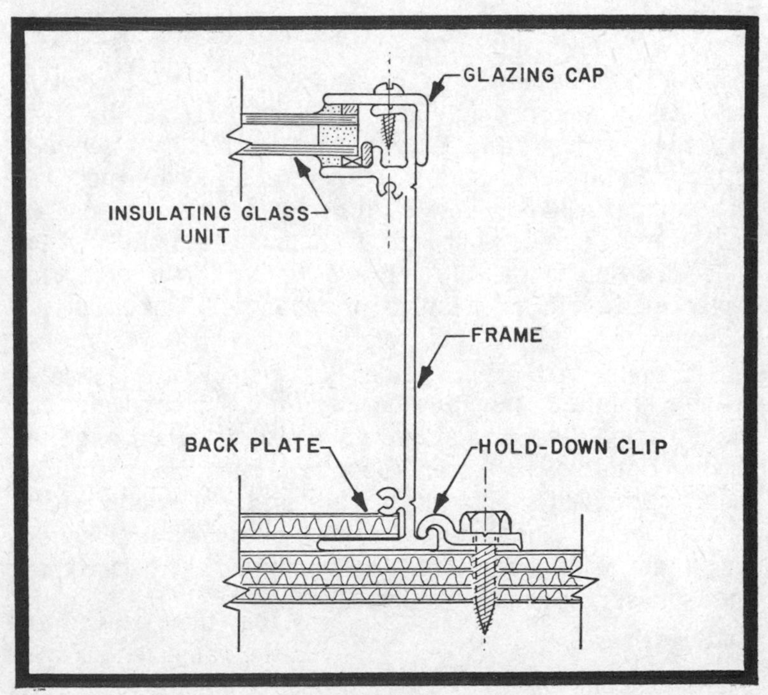

Fig. 1-16. Typical method of glazing and attachment using the Alcoa framing system.

Table 1-1. Spacing and Clearances for Glazing Methods Shown in Fig. 1-17.

Detail	Glass Type and Thickness	Edge Clearance A	Nominal Cover B	Setting Block	Gasket Spacing C
face glazing wood sash	3/16″ double strength	3/16″	5/16″	none	1/8″
glazing cap metal sash single glazing	3/16″ tempered	1/4″	3/8″	yes	3/16″
glazing cap metal sash dual glazing	1/8″ tempered lights with 1/2″ air space	1/8″	1/2″	yes	1/8″

Glazing Techniques

Three common glazing details are shown in Fig. 1-17. Table 1-1 provides dimensions for items depicted in Fig. 1-17, including recommended edge clearances A, nominal cover B, and gasket spacing C for each detail using various types and thicknesses of glass. The C dimension is also the width of the setting block (which is illustrated by a square with diagonal lines drawn through it in Fig. 1-17). Setting blocks are usually made of short pieces of wood or neoprene and serve to position the glass in the opening. They are spaced around the glass edge to prevent shifting. When purchasing factory assembled insulating glass units, the manufacturer's recommendations on spacing should be followed. Although face glazing (see Fig. 1-17) offers a simple approach, I would not recommend it for collectors due to the difficulty in removing the cover for replacement, interior cleaning, or repairing a leak inside the collector.

As mentioned earlier, fiberglass-reinforced plastics, if suitably stabilized, may also be used for collector covers. The same glazing details as shown previously for glass might be used, but additional clearances for expansion and contraction must be provided if these plastics are used. For example, 0.040 inch thick fiberglass-reinforced plastic expands in thickness, length, and width about four times as much as 1/8-inch thick sheet glass for the same temperature range.

Cutting Glass

It is not possible to cut glass after it has been tempered. But if you select double-strength window glass as a single glazing, or if you elect to make your own insulating glass using

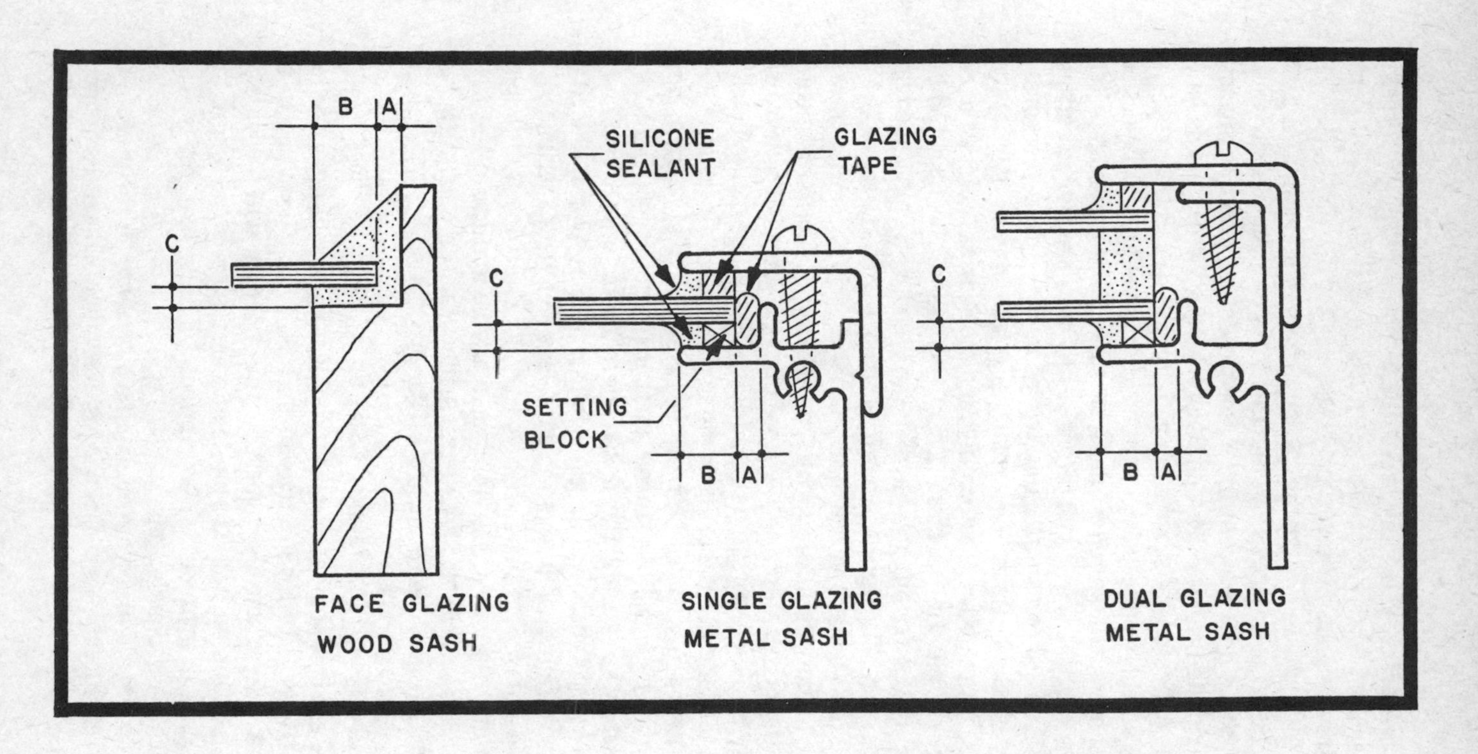

Fig. 1-17. Common glazing details showing edge clearance (A), nominal cover (B), and gasket spacing (C).

this material, you may want to do some cutting yourself, rather than having it done at the glass supply house. If so, here are some basics on how to cut glass.

You will need an inexpensive hand-held cutter, a flat surface large enough to handle the glass size you plan to work with, some light penetrating oil or turpentine, and a hardwood straightedge (about 1/4 inch thick by 3 inch wide, and as long as the collector you plan to build).

Pad the work surface with an old carpet or with layers of newspaper, placing the glass on top of the padding. After measuring the size, swab the oil or turpentine along the cutting line. Then, with a smooth but firm pressure on the cutter, draw it across the glass in *one stroke*. All you need to do is *score* the glass. If the cutter is accomplishing the scoring, don't go back over the line. Only retrace a spot if it has not been scored by the cutter. Pressing too hard on the cutter will chip the edges of the glass. Too light a pressure will not result in a score.

When breaking the glass after scoring, remember to always snap the piece *away* from the score line. This means that the break *starts* at the score mark. Figure 1-18 shows two methods of snapping the scored edge. For either method, don't wait too long after scoring to make the break, and use a quick, downward pressure. Narrow strips (less than 2 inch) are difficult to break by these methods; professional glaziers generally use special flat-nosed pliers for this type of cut.

Heavy sheet or plate glass in thicknesses over 3/16 inch is more difficult to cut evenly, so unless you are experienced, I would not recommend that you try it. Even for the double-strength window glass (which is available in thicknesses 0.115 to 0.134 inch), you should try some practice cuts on scrap pieces before attempting any larger production.

Construction of Insulating Glass Units

Factory-fabricated insulating glass units are suggested for use. However, if you have special skills and want to try to make the assembly yourself, this section will provide a construction outline. If you decide assembly is too tough, the section still provides a description of quality features to look for when purchasing a unit.

Materials and workmanship will determine the useful life of a sealed glass unit. Once moisture vapor penetrates the seal at a level that can no longer be adsorbed by the *desiccant*,

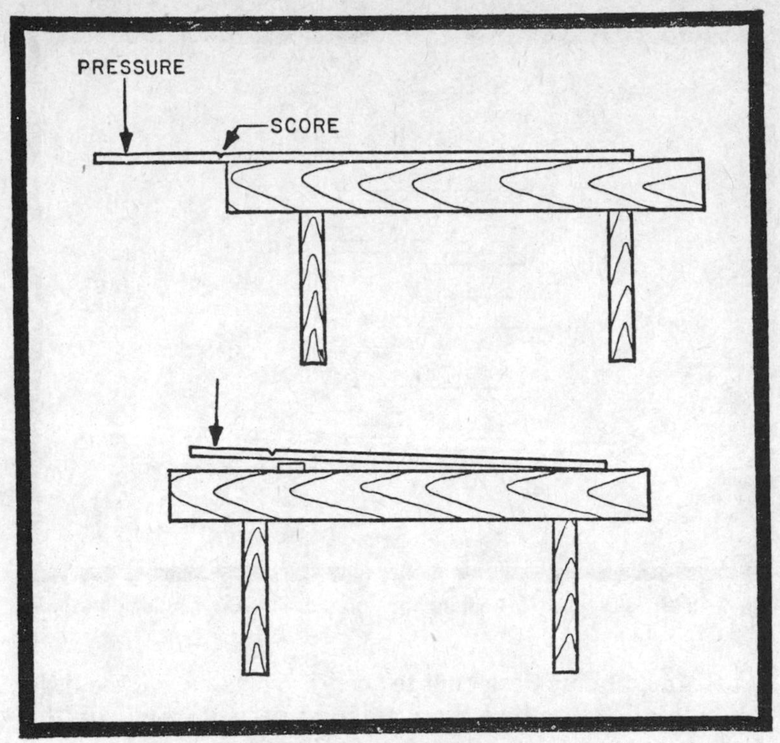

Fig. 1-18. Two methods of breaking a glass edge after scoring.

condensation forms between the glass covers, and the unit must be replaced. (A desiccant is a substance that has a high affinity for water and is used to absorb moisture.) Here are the key factors influencing the life of the unit:

- type of sealants
- type and quantity of desiccant
- design and cleanliness of spacers
- correct corner treatment
- cleanliness and flatness of glass
- quality of workmanship

First, let's discuss a single-seal system, as shown in Fig. 1-19. The air spacer (either aluminum or galvanized steel) separating the glass has good design features. The metal edge that the glass rests on is only wide enough to lay flat on the glass without twisting. This provides a *longer* sealant line and greater resistance to moisture vapor. Also note that the spacer frame has a good tight seam—enough space to allow the tube

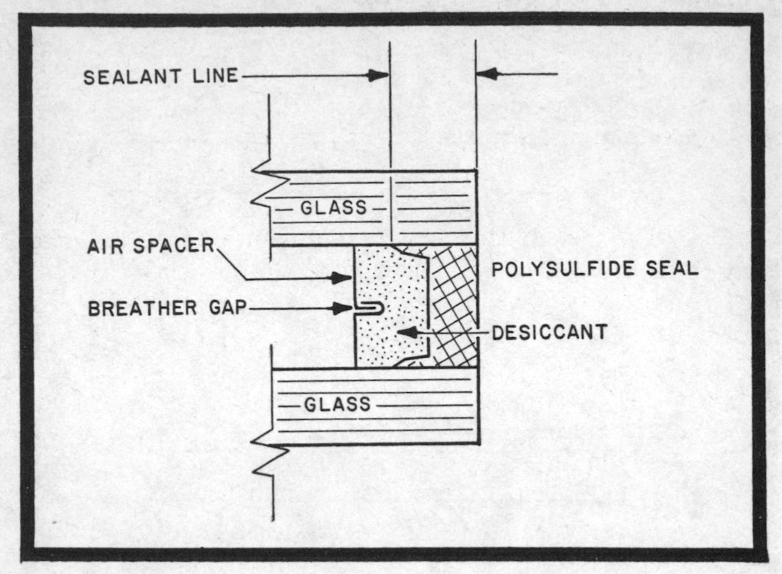

Fig. 1-19. Cross section of an insulating glass unit with a single polysulfide seal.

to breathe, and its desiccant to adsorb moisture, but too little space to allow the dew point to build up and inactivate the desiccant before the frame is installed on the glass.

On single-seal systems, corner keys may be used in the assembly of spacer frames. These are generally zinc die

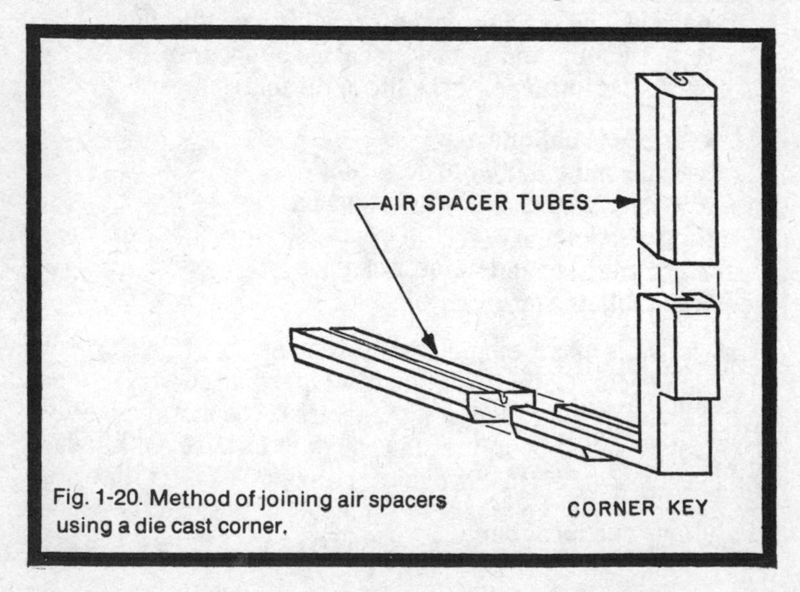

Fig. 1-20. Method of joining air spacers using a die cast corner.

castings, and they slip inside the tubing to form a 90° corner. Figure 1-20 shows this construction. With the corner key closing one end of the tube, the cavity is then filled with a desiccant. Many manufacturers use a 50 percent silica-gel and 50 percent molecular-sieve mixture to fill the cavity. This provides a good capacity to adsorb both water vapor and solvent vapors from sealants. The other sides of the frame are then assembled, each one being filled with the desiccant. Once the frames are full of desiccant, they must be sealed in the glass before one hour has elapsed. Both the frame and glass must be perfectly clean before assembly. Any oil or contamination will prevent the sealant from adhering properly, so I would suggest wearing clean, white cotton gloves when touching these surfaces.

The sealant generally used for single-seal systems is polysulfide. This material is available in one-component or two-component types. The latter type requires mixing. The sealant is either extruded into the cavity using a caulking gun, or troweled into place. In either case, there should be no misses, gaps, or air bubbles around the entire perimeter of the unit.

Double-seal systems, when properly constructed, can provide a life expectancy of 20 years or more. Although they are beyond the scope of the average do-it-yourselfer, I will describe a typical assembly so you can get a feel for the construction.

The double seal of Fig. 1-21 has a frame with either soldered aluminum or brazed galvanized steel corners. The cast corner keys used on single-seal systems should be avoided because in most cases this is where leaks occur. Also, note that a butyl seal is applied first *between* the spacer frame and the glass, with the remaining cavity being filled by polysulfide—hence the name, *double seal.* In Chapter 2, I will list some of the standard sizes of double-seal units available, together with their approximate cost.

PLUMBING AND ELECTRICAL CONSIDERATIONS

Demands on pumps, valves, collectors, controls, and storage tanks will vary with the individual system chosen, but you must have a basic knowledge of the factors that go into their selection in order to make modifications for custom installations and local building codes.

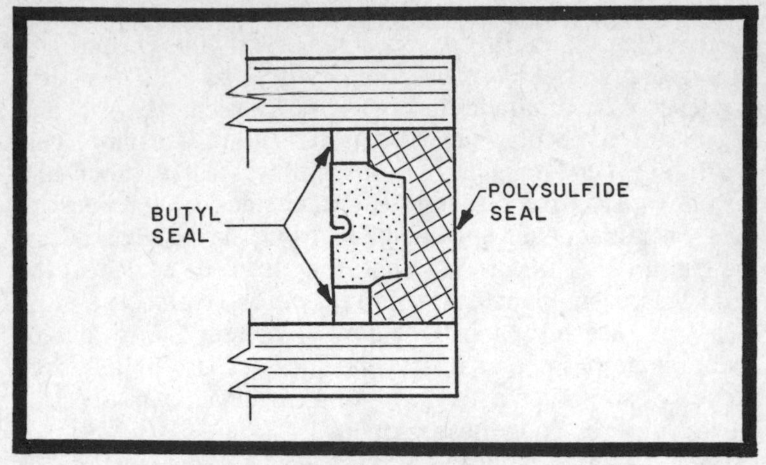

Fig. 1-21. Cross section of an insulating glass unit that is double sealed with butyl and polysulfide.

Plumbing

All the systems in this book are *liquid systems*; that is, ones in which a fluid, rather than air, is circulating through the collector to transfer heat. In dealing with fluids, the following factors must be considered:

- pressure
- flow rate
- head
- expansion
- temperature
- toxicity
- flammability
- freezing point
- chemical compatibility

Some systems operate at water *line pressure*, requiring that all materials used to contain the water, including joints, be strong enough to handle this pressure without leaking or rupturing. Line pressure will vary in different localities but is generally assumed to be a maximum 100 pounds per square inch. Where local water pressure is in excess of this value, building codes have special requirements.

As fluids increase in temperature, they *expand* in volume. If they are contained in tanks and pipes that do not permit volume expansion, the *pressure* increases. If safety devices were not required by building codes, pressurized lines or tanks could explode with catastrophic results. For this reason, all water-heating appliances which are installed in a closed system of water piping, or any water heaters connected to a separate storage tank with valves between the heater and tank, must be provided with a water pressure-relief valve. Each pressure-relief valve must be code-approved, automatic, and set to open when pressure reaches 125 pounds per square inch.

In addition to the primary temperature control or *thermostat* on a water heater, a temperature limiting device must also be installed. Combination temperature/pressure valves, as pictured in Fig. 1-22, are available commercially. These are commonly called *TP valves*. Precautions are necessary at the exit or *drain* side of these valves to prevent injury or damage from scalding water when the valve opens. Therefore, codes will require that a drain line be attached to the valve and that the line terminate in a safe place.

When potable water is circulated through the collectors in a system, as pictured in Fig. 1-3, a separate *expansion* tank is not necessary. Raising the temperature of water from a supply line temperature of 50°F to 200°F causes the *volume* to increase by 4 percent, but rarely will this exceed the capacity of the storage tank. If it does, the pressure-relief valve will open and discharge some water. On the other hand, if a glycol and water mixture is used to prevent freeze-ups in a *closed* system, a 6 percent volume expansion will occur over the same temperature range. Because glycol is expensive, an expansion tank will be used to collect the overflow. This same principle is used on some automobile cooling systems, with a clear plastic *expansion* tank placed next to the radiator to catch the overflow.

Circulating tap water through the collectors is suitable for mild areas where freezing weather does not occur. Using copper plumbing, this concept eliminates the problems of *toxicity*, *flammability*, and *chemical incompatibility*. These factors become important, however, when glycols or other heat-transfer fluids are specified in the design.

Flow rate and *head* are the last two variables I want to introduce here. The *rate* at which water flows through the collectors affects their *efficiency*. *Slow flow* produces higher temperatures but takes longer to heat the storage tank; it also causes efficiency to decline. A *high flow* rate may not provide

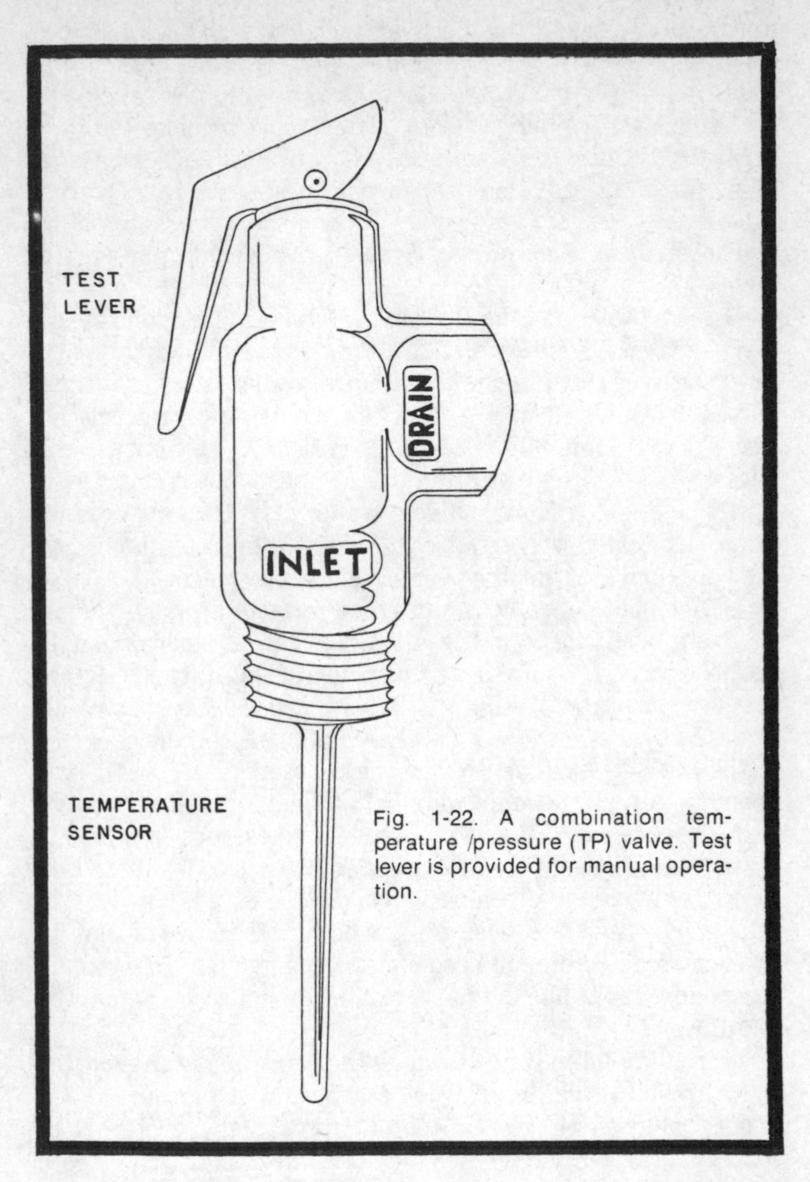

Fig. 1-22. A combination temperature /pressure (TP) valve. Test lever is provided for manual operation.

a sufficient temperature gain to meet the system requirement. For example, a solar swimming-pool heater is designed to operate at a high flow rate, raising the large volume of the pool water perhaps 10° F. With a slow flow in this application, it would be like trying to heat the pool with a candle. The flame from the candle is indeed hot, but it would take forever to heat the pool itself. There is an *ideal* flow rate for each application, but this rate varies with changing weather conditions generally an average value is established and the flow is *fixed*.

The circulation pump controls the *rate* of fluid flow which is affected by the *head*, or pressure, of water it must pump against. There are frictional forces resulting from valves, fittings, and the pipe or tubing used, plus the vertical height to which the water must be raised by the pump. The total *head* is determined when these factors are all added together. Performance curves for pumps show their head capability in feet of water for various flow rates. For example, the pump I use in the Chapter 7 system is the Grundfos model UP 25-42 SF. The curve for this unit is shown in the manufacturer's literature in Chapter 3. In Chapter 7 I will explain how to set the flow rate for your specific conditions.

Electrical Connections

In general, the electrical devices in a solar system are limited to the differential thermostat with its temperature sensors, pump(s), and a solenoid valve. The thermostat and the pump operate on an input of 120 volts, 60 cycles. Connections are not difficult, but a few basic considerations must be observed.

- Wire size must be large enough to handle the load.
- Appliances should be grounded to prevent electrical shock.
- Every switch, joint, and electrical outlet must be contained inside a *box* (generally made of metal) designed for such purpose, and covered.
- Electrical cable must be securely anchored to junction boxes.
- An individual circuit (serving no other load) should be provided for systems having an automatically started motor (pump, furnace blower, etc.).
- Black wires are always *hot* and are the only wires connected to switches controlling 120-volt electrical devices.
- White wires are always *neutral* or grounded and must never be fused, switched, or interrupted in any way.
- Wiring should be tested for continuity before connecting to house wiring or live circuits.

Selecting the correct diameter or *size* of wire is necessary to insure that the proper voltage reaches the appliance and

Table 1-2. Selection of Type T Wire Sizes for 115-Volt Appliances.

Wire Size	Diameter Inches	Ampacity	Watts at 115 Volts
14	0.064	15	1725
12	0.081	20	2300
10	0.102	30	3450

that the wiring does not overheat. Type T copper wire is the type most commonly used. It is a solid copper conductor with a layer of plastic insulation. The current-carrying capacity (*ampacity*) and the wattage for three common wire sizes are shown in Table 1-2. The *wattage* (voltage multiplied by amperage) of each appliance is shown on its nameplate. Total the requirements for the system you are considering, and make sure that the wire size selected will handle the load. Type NM (nonmetallic, unsheathed) cable is frequently used to connect appliances; it is easier to work with and less expensive than rigid conduit or armored cable. Type NM cable, with ground, consists of two type T conductors (a black and a white insulated wire) with an uninsulated wire of the same size all contained within a cable covered with braided fabric. This is called Romex cable in the trade, and is pictured in Fig. 1-23. For damp locations and outdoor use, type NMC cable must be specified; it has a weatherproof plastic covering instead of the paper and fabric used on type NM cable. I prefer type NMC around pumps or plumbing lines where there is a possibility of leakage. In either case, I recommend that you purchase the cable with a ground wire included.

Wire selection and electrical connections may vary to satisfy local code requirements, but there is a basic method of making connections using NMC cable and metal outlet boxes. Strip the cable back, as shown in Fig. 1-23, using a pocket knife or wire stripper. Be careful not to damage the insulation. Knock out the desired openings in the metal outlet box and install a cable connector of the type shown in Fig. 1-24. The connector on the left of the figure is best suited for flat cable and straight connections. The connector on the right is designed for larger diameter cable and sharp 90° bends. The connector is fastened to the outlet box with the nut provided. Insert the NMC cable into the connector until the outside wrap is just visible inside the box, and tighten the clamp screws.

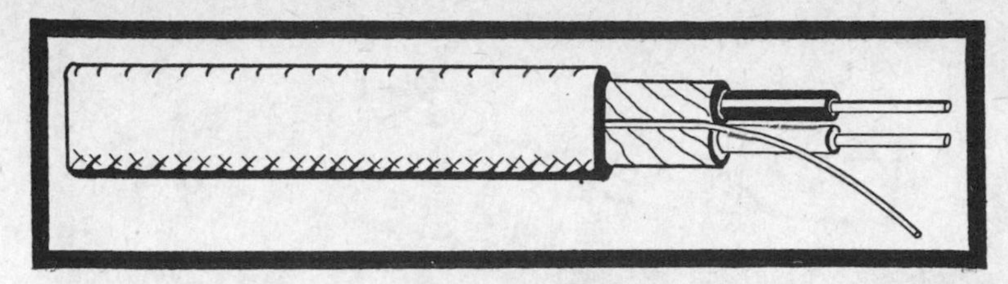

Fig. 1-23. Construction of nonmetallic cable (NM). A ground wire and two insulated conductors are included.

Wire-to-wire connections are made inside the box using solderless connectors or *wire nuts*. These have tapering threads inside and a plastic outer shell. They are screwed onto the wires. Figure 1-25 shows a cross section of the connector. The bare ground wire should be connected to the outlet box either with a screw or grounding clip which snaps over the edge of the box. Grounding screws on appliances or receptacles are usually green.

A simple method of testing wiring requires a doorbell and two 1 1/2-volt dry cell batteries hooked in series. *With the power off*, connect the cells to the black and white wires at the *origin* of the line to be tested. Similarly, attach the doorbell at the *end* of the line, and if there are any switches in between, be sure they are turned on. If the bell rings, the circuit is complete.

Now check for grounding by connecting the bell between the black wire and the metal outlet boxes at each location. A ringing bell confirms the ground. If you have trouble getting a

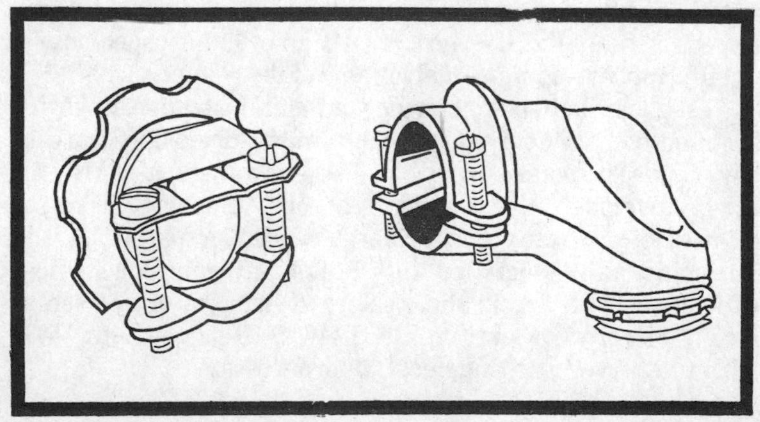

Fig. 1-24. Two types of cable connectors. Type on the left is suggested for flat NM cable in straight runs.

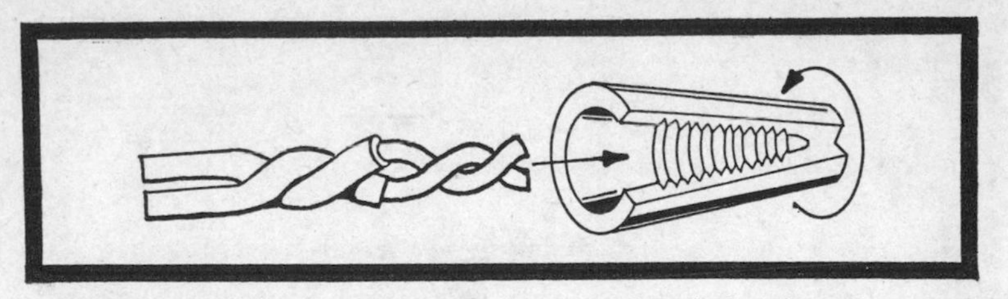

Fig. 1-25. Solderless connector, commonly called a wire nut.

ring, recheck all connections and test each section of line individually.

OBTAINING A BUILDING PERMIT

A solar heating system should not be installed before you obtain a permit from your administrative authority. This may be called the Building Department, Department of Building and Safety, or something similar, but regardless of the name, the department is there to enforce building codes designed to protect your health and safety. Approval must be obtained from this department before any work is started.

You may be required to submit a plumbing, electrical, and perhaps structural plan to the department for review and for a check of compliance with the applicable codes. If so, plans are usually required in duplicate. Once the plans are approved, a nominal fee will be charged and a permit issued. As the work commences, one or more inspections by the department are usually required. It is your responsibility to notify the department when you are ready for inspection (generally a minimum of 24 hours in advance of when you want the inspection made). If the work is satisfactory, the inspector will usually approve or *sign off* all or part of the permit.

I suggest that you telephone your administrative authority first and determine what its requirements are before making any plans. The solar energy field is so new that codes are still being developed. I am aware of only one such code in print—the Uniform Solar Energy Code, prepared by the International Association of Plumbing and Mechanical Officials. If you are in the western states, you may wish to obtain a copy by writing to IAMPO Headquarters, 5032 Alhambra Ave., Los Angeles, California 90032.

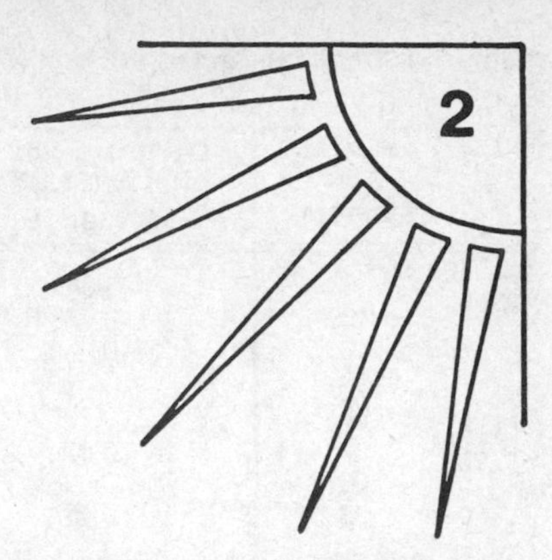

Materials: Terminology, Price, and Sources of Supply

The building and construction industry employs people of many different trades—plumbers, carpenters, sheet-metal workers, glaziers, etc. Over the years, each trade has adopted its own terminology and methods of measurement that make standardization difficult. In addition to providing an approximate price range for preliminary cost estimates, this chapter will explain the terms that are common throughout the industry. If you know *how* to ask for the parts, the job becomes much easier.

ALUMINUM

Both extruded and sheet aluminum play a part in the manufacture of solar collectors. The dictionary defines *extrude* as "to push or force out." For example, when you squeeze a toothpaste tube, you are *extruding* paste through the round opening at the mouth of the tube. In metalworking, the principle is the same, but the round opening is generally replaced by a die having a more useful shape, such as a channel, angle, or custom collector frame. Aluminum extrudes more rapidly than brass, copper, or steel, thereby permitting more flexibility in design at lower cost. An example of this is the integral screw slot explained in Chapter 1. Over 90 percent of the aluminum extrusions used in the construction industry are made from *alloy 6063 in the T5 temper*; this is what you should order for the collector frames. *Structural*

Table 2-1. Decimal Equivalent and Nominal Weight of Aluminum Sheet for Various Brown and Sharp Gauges.

Gauge No.	Decimal Equivalent of Brown & Sharp (B&S) Gauge	Nominal Weight of Aluminum Sheet lb/sq ft
14	0.064	0.922
16	0.051	0.734
18	0.040	0.576
20	0.032	0.461
22	0.025	0.360
24	0.020	0.288
26	0.016	0.230
28	0.013	0.187
30	0.010	0.144

extrusions are commonly used for items like mounting stands, and these angles and channels may be stocked by metals distributors in *alloy 6061 in the T6 temper*. These offer higher strength and sell for a slight premium. Extrusions are normally sold on a price-per-pound basis, so you need to know the weight of the shape to determine the price.

When it comes to aluminum sheet, any of the *common alloys*, 1100, 3003, or 5005, may be considered, depending on availability. The thickness of aluminum sheet is generally specified in thousandths of an inch (0.032 inch, etc.). You may still hear people refer to aluminum by gauge number (such as 20 gauge, 14 gauge, etc.); these figures relate to the Brown and Sharp gauge system for copper sheet. Table 2-1 shows a comparison of the systems, and also the nominal weight per square foot of aluminum. This information is helpful in calculating the price.

All pricing in this chapter is approximate and based upon the range you could expect to pay, from the sources shown, as of this writing. There are *extras*, such as cutting charges, that must be added whenever you ask for a small size that has to be cut from a standard piece. These small, odd pieces are called remnants or *rems*. Sometimes you can locate rems that will suit your need and save money. Metals distributors or service centers generally have a customer service or cash sale desk to help you in your selection. Table 2-2 lists prices of typical Alcoa solar extrusions stocked by distributors.

COPPER

The roofing trade has used copper sheet for years for both the roofing surface and flashings. It is customary to relate the

Table 2-2. Typical Prices of Alcoa Solar Extrusions from Distributors. February 9, 1977.

	Die Number and Part Name		
	459831 Frame	459841 Glazing Cap	459851 Hold-Down Clip
Length Stocked	20′	20′	20′
Pounds per Length	12.72	2.50	4.12
Quantity Ordered	Price per Foot (in Mill Finish)		
under 25	$1.025	$0.200	$0.329
25–49	0.897	0.175	0.287
50–99	0.808	0.158	0.259
100–199	0.751	0.146	0.240
200–299	0.719	0.140	0.230
300–499	0.707	0.138	0.226
500–999	0.624	0.121	0.199
1000–1999	0.598	0.116	0.191
2000 and over	0.586	0.114	0.186

Courtesy:
Southern Aluminum Finishing Co.
1581 Huber Street N.W.
Atlanta, Georgia 30318

thickness of the material to its *weight* per square foot. For example, in Table 2-3 you can see that *16-ounce* copper is actually 0.022 inch thick. For solar collector absorber plates, the *16-ounce* copper used for roofing material is not required. Ten-ounce copper sheet will provide good heat transfer when the collector tubes are spaced 6 inches apart, and it is considerably cheaper. If you can locate the 10-ounce material, you should use it for the designs in this book.

Table 2-3. Nominal weight of copper sheet for various thicknesses.

Nominal Weight Of Copper Sheet oz/sq ft	Thickness in inches
8 oz	0.011
10 oz	0.014
14 oz	0.019
16 oz	0.022
20 oz	0.027
24 oz	0.032
32 oz	0.043

Table 2-4. Dimensions for Selected Sizes of Copper Water Tube.

	Nominal Size Inches	Outside Diameter (O.D.)	Wall Thickness	Weight lb/ft
TYPE K	1/4	.375	.035	.145
	3/8	.500	.049	.269
	1/2	.625	.049	.344
	5/8	.750	.049	.418
	3/4	.875	.065	.641
	1	1.125	.065	.839
	1 1/4	1.375	.065	1.04
	1 1/2	1.625	.072	1.36
TYPE L	1/4	.375	.030	.126
	3/8	.500	.035	.198
	1/2	.625	.040	.285
	5/8	.750	.042	.362
	3/4	.875	.045	.455
	1	1.125	.050	.655
	1 1/4	1.375	.055	.884
	1 1/2	1.625	.060	1.14
TYPE M	3/8	.500	.025	.145
	1/2	.625	.028	.204
	3/4	.875	.032	.328
	1	1.125	.035	.465
	1 1/4	1.375	.042	.682
	1 1/2	1.625	.049	.940
DWV	1 1/4	1.375	.040	.650
	1 1/2	1.625	.042	.809

Another substitution that may be made is to use 6-inch wide copper strip in coil form to replace the sheet. If you can locate the material, soldering the 6-inch strips along the full length of the riser tubes (3-inches on each side of the tube), will work out. It is not necessary to solder the strips together, but they should be close to each other at the edges to intercept a maximum amount of sunshine. The problem I had was in locating strip copper. Most roofing and sheet-metal shops now use *aluminum* flashing because it is much cheaper and it does the job.

Copper plumbing components have been used for years with good success, and they are universally accepted by the building codes for carrying potable water. For this reason, I used copper components in my designs.

From a *cost* standpoint, aluminum sheets soldered to copper tubes would provide the best value for solar collector absorbers—but the soldering job is just too difficult to recommend for do-it-yourself projects.

Copper tube is available in four basic types with varying wall thickness. Type M is the most economical and is satisfactory for low pressure service. Type L has a heavier wall and is perhaps the most common in residential plumbing systems. Type K which is specified for systems with extra-high pressure or underground lines. Lastly, type DWV is for drainage, waste, and vent lines. It has the lightest wall thickness and is not designed for pressure applications. Table 2-4 provides dimensions for common sizes of copper tube.

Each type of copper water tube has a safe internal working pressure which varies with the service temperature of the water it carries. Other factors to be considered are the strength of the fitting, the solder, and the soldering techniques. Local codes vary as to the requirement for wall thickness. I chose type M due to its lower cost, but as you can see from Table 2-5, type K will handle almost *twice* the pressure of type M, and would stand up better if water should freeze in the lines. Depending on your local conditions, the heavier grade may be a bargain.

Most hardware stores will carry 1/2-inch copper tube. It is the most popular size and is frequently available at sale prices, so watch for bargains. Many of the other tube and fitting sizes specified must be acquired from plumbing distributors because they are not commonly used. If the retail distributor does not have the item you need in stock, he can order it from the wholesaler; your order should be of sufficient size to encourage him to do this. Table 2-6 lists typical retail tubing prices.

Table 2-5. Copper Tube Safe Working Internal Pressures

Nominal Size In Inches	Outside Diameter	Type K 150°F	Type K 250°F	Type L 150°F	Type L 250°F	Type M 150°F	Type M 250°F
	1/8	3130	3030				
	3/16	1950	1890				
1/8	1/4	1530	1480	1160	1120	1160	1120
	5/16	1200	1160				
1/4	3/8	1200	1160	900	870	750	730
3/8	1/2	1170	1130	800	770	560	540
1/2	5/8	920	890	740	720	500	480
5/8	3/4	760	730	650	630	450	430
3/4	7/8	880	850	590	570	400	390
1	1 1/8	680	660	510	490	340	330
1 1/4	1 3/8	550	530	460	440	340	330
1 1/2	1 5/8	520	500	430	420	340	330

Table 2-6. Approximate Price Per Foot for Copper Water Tube.

Nominal Size (Inches)	Tube Type		
	K	L	M
1/4	$.30	$.27	$.23
3/8	.53	.40	.30
1/2	.67	.56	.40
5/8	.84	.72	.55
3/4	1.23	.89	.65
1	1.63	1.28	.95
1 1/4	2.08	1.74	1.36
1 1/2	2.69	2.25	1.85

Copper plumbing fittings are available in so many different types and configurations that confusion can easily result. For this reason, I am including several figures showing cross sections of parts specified in the designs, with accompanying terminology. There are two basic classes of pressure fittings for water lines having soldered connections—cast and wrought. Either type will work for the applications in this book, and you should select according to price and availability.

For soldered joints, the fitting is designed to slip over the *outside* of the tube with enough clearance to allow solder to flow into the joint. There is a ridge or ledge inside the fitting which stops the tube at a specified distance. You can see this in the cross-sectional drawings. As the fitting accepts the outside diameter of the tube, one size will work regardless of the wall thickness chosen. For example, a 1/2-inch fitting will work on 1/2-inch nominal size type K, L, and M water tube. Fittings designed to be soldered to copper lines have a C designation. If

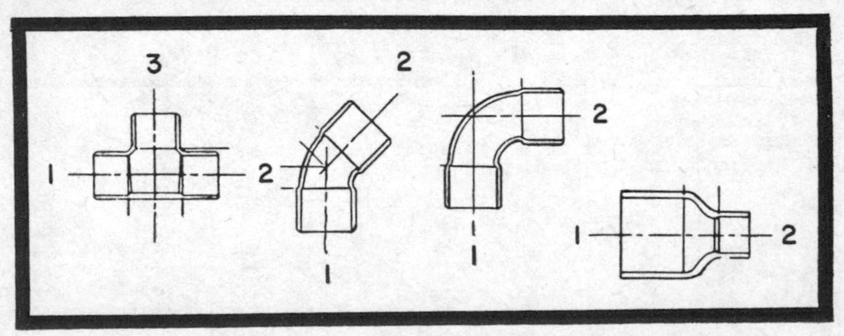

Fig. 2-1. Types of fittings used for soldered connections.

the joint is copper-to-copper, then it is called a C × C fitting. If there is to be a 3-way soldered connection, such as a tee, then C × C × C is specified. Figure 2-1 shows several fitting types for soldered connections. Note that in Fig. 2-1 there are numbers shown on the centerline(s) of the parts. These refer to the dimensions of the particular opening and the *order* in which these dimensions are written. For example, a tee to connect 1/2-inch risers to a 1-inch header pipe (used in all my designs) is specified 1 × 1 × 1/2, corresponding to the numbers 1, 2, and 3, respectively, on the drawing. The *third* opening is the 1/2-inch size. M and F stand for male and female *threaded* connections, and six types of these connectors are shown in Fig. 2-2. The 90° union ell (short for elbow) is a handy fitting. A union consists of two machined fittings which are joined by a threaded coupling. Unscrewing the coupling allows disassembly of the parts. Unions are used to connect lines to components which may require service or replacement. I use them to connect the collectors and water-storage tank. The union shown in Fig. 2-2 is of the C × F, or copper to female, type. Table 2-7 lists typical retail prices for fittings used in the solar projects.

GLASS AND GLAZING MATERIALS

Glass has been used for many years as a building material, and specifications, codes, and glazing techniques are

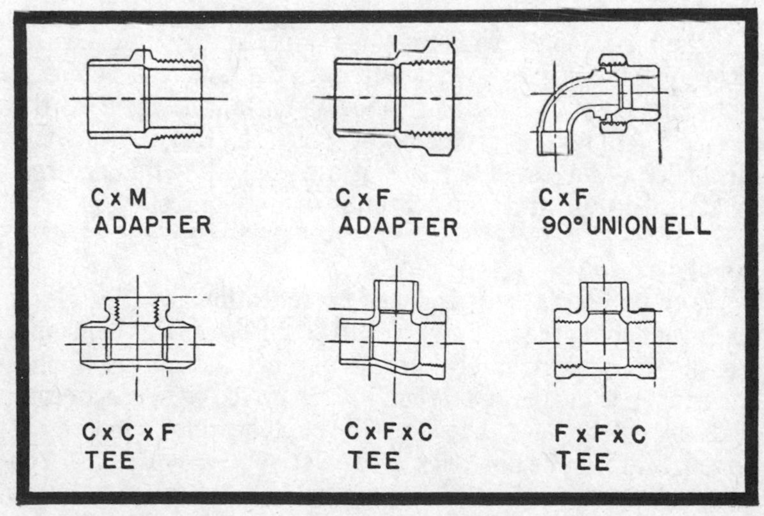

Fig. 2-2. Connectors for adapting copper pipe to threaded fittings.

Table 2-7. Typical Retail Prices for Fittings. January 15, 1977.

Size (Inches)	Fitting Type	Description	Price per Piece
1/2	CPVC plastic	unthreaded coupling	$0.27
3/4	CPVC plastic	unthreaded coupling	0.18
3/4×3/4	C×M copper	adapter	0.37
1×3/4	C×M copper	adapter	1.45
3/4×3/4×1/2	C×C×F copper	tee	0.70
3/4×3/4×3/4	C×C×F copper	tee	1.25
3/4×3/4×3/4	C×F×C copper	tee	1.25
1×1×1/2	C×C×C copper	tee	1.30
1×1×3/4	C×C×C copper	tee	1.30
3/4×1/4	C×F copper	90° ell	1.60
3/4×3/4	C×F copper	90° ell	1.39
1×1/2	C×C copper	90° ell	1.35
3/4×3/4	C×C copper	union	1.68

well established. There are many types of glass available, such as:

- window and sheet glass
- plate or float glass
- tempered or heat-strengthened glass
- pattern glass

But for the solar collector designs in this book, only three types are considered: double-strength window glass, tempered glass, and insulating glass.

Double-Strength Glass

Double-strength window glass is commonly produced by drawing molten glass vertically out of a tank to a height of approximately 30 feet and, once it cools, cutting it to the desired length. As you might suspect, there are some distortions (waves or lines) that are caused by this process, but these do not affect the operation of the solar collector.

Tempered Glass

Tempered glass is produced by reheating annealed glass, then quenching (or chilling) it with an air blast. This technique results in an extremely strong product insofar as impact resistance is concerned. When it does break, it does not crack like annealed glass, but literally explodes into hundreds of small, marble-like particles with relatively smooth edges. You can rub a handful of these particles between your hands without getting cut. Tempered glass is generally used for automobile windshields and is specified for sliding patio doors due to its strength and safety features. Because tempered glass cannot be cut after processing, you must buy it to size. Collectors are generally made to accommodate the standard sizes.

Insulating Glass

Insulating glass, by my description, is a sandwich made up of two glass panes separated by a metal spacer. A desiccant is placed inside the spacer to *adsorb* water and solvent vapors, and the edges of the glass are then sealed. Figure 1-19 showed this assembly. In some factory-assembled units, the air trapped between the glass panes is dehydrated and the edges double sealed to insure long life. A complete discussion of the pros and cons of using *dual glazing* (insulating glass) on solar collectors is provided in the *Homeowner's Guide to Solar Heating & Cooling*, TAB Book No. 906. But in summary, double-pane collectors are best suited to systems in colder, northern climates. They use a heat exchanger in the collector loop and an absorber painted with flat-black paint (nonselective).

"Low Iron" Glass

The amount of iron oxide in glass affects its *transmission*, or the percentage of solar energy that will pass through it. The *higher* the iron content, the *lower* the solar transmission. You can determine the approximate degree to which glass exhibits this quality by looking at the edges of the glass: the greener the color, the higher the iron content. The best grade glass, somctimes referred to as *water white*, will transmit a maximum of 91 percent of the energy; ordinary glass would be in the 85 percent range. For projects in this book, I would suggest that you avoid paying a premium for low-iron glass, but if you have a choice, select the brand with the whitest edges.

Glass Terminology

Those in the trade will not generally refer to a *piece of glass*; instead they call it a *light* (one light equals one piece, etc.). An exception is insulating glass which is made using two lights, and is called an insulating glass *unit*. Other terms used are *sealant, glazing tape, desiccant,* and *setting block*. You may wish to refer to the glazing section of Chapter 1 which supplements the descriptions that follow.

A *sealant* or sealing compound is used to fill and seal a joint or opening. *Wet* sealants generally come in a cartridge and are applied with a *caulking gun*. *Glazing tape* is also available in a roll with either a flat or round bead shape and is applied by placing the material by hand. There are three classes of sealants generally used in solar collector work:

- butyls
- polysulfides
- silicones

The use of butyl and polysulfide sealants in the construction of insulating glass units was covered in Chapter 1. Each material has advantages for specific applications. Butyls have only fair weather resistance, while polysulfides are rated very good in this category. For this reason, a polysulfide is used *outside* the butyl on a dual-seal insulating glass unit. Silicones provide excellent weatherability and elasticity. This is of special benefit where you have expansion problems to consider with fiberglass-reinforced plastic covers.

One last word on sealants. Each product has its advantages, but the maximum recommended service temperature should be considered:

Sealing Compound	Service Temperature Range
Butyls	−20° to 180° F
Polysulfides	−40° to 200° F
Silicones	−60° to 250° F

These are the general guidelines, but the manufacturer's specifications should be the determining factor in the selection.

Table 2-8 provides typical retail prices on glazing materials which may be purchased at a glass supply house. I use only tempered glass. Because this material may not be cut after factory tempering, standard sizes are shown in the table; but 1/8-inch and 5/32-inch thicknesses may be difficult to find. Also, the 46-inch width is generally not stocked locally. Your glass supplier can order it, though.

Another suggestion that will save money is to buy *used* glass if you can find it. For example, many glass houses will disassemble an insulating unit that may be damaged. The

Table 2-8. Typical Retail Prices for Glazing Materials. January 15, 1977.

Item Description	Size	Price	
		Small Quantity	Case Lots
tempered glass	1/8×34×76	$1.90/sq ft	$1.43/sq ft
	3/16×34×76	1.96	1.77
	3/16×46×76	—	1.86
	3/16×46×96	—	1.86
insulating glass units (1/2" air space)	1/8×34×76	76.80 each	56.95 each
	3/16×46×76	102.16	85.04
"Tedlar" PVF film	0.004 in. thickness	0.59/sq ft	0.43/sq ft
400×RB 160SE fiberglass reinforced plastic	.040 in. thickness	0.81/sq ft	0.50/sq ft
silicone sealant	1/12 gal cartridge	5.30 each	3.80 each
glazing tape	1/4"×1/4" (50-ft roll)	3.00/roll	2.00/roll

lights that are salvaged may have some sealant on the edges, but they are easily cleaned. I obtained a 30 percent discount by buying this way.

FASTENERS

I will not go into great detail on fasteners other than to stress that only corrosion-resistant materials be used—aluminun, cadmium-plated steel, or stainless steel. Cadmium-plated steel is carried by most hardware stores and is not expensive.

The only fastener that may need some description is the sheet-metal screws specified for assembly of collector frames and glazing caps. Two types of sheet-metal screws may be considered—the conventional kind used for relatively thin gauges of material, and the self-tapping variety which I recommend for assembly of aluminum extrusions. These are both *type A* screws which have a sharp point. The collector designs in this book use a No. 10 size which means that the major diameter is approximately 0.188 inch. The Alcoa framing system is made with screw slots that are 0.142 inch in diameter, so the screws will bite into the metal as they are tightened. The holes drilled into the glazing cap and frame

Table 2-9. Insulating Value of Various Materials in a One Inch Thickness.

MATERIAL	K FACTOR	R VALUE
Urethane foam	0.14	7.14
Fiberglas	0.24	4.17
Polystyrene	0.26	3.85
Foamed rubber	0.28	3.57
Foamed glass	0.39	2.56

ends are 0.188 inch in diameter (a 3/16 inch drill is used), so the screws will pass through.

INSULATION

Tubular insulation is required for plumbing lines, and fiberglass semi-rigid board is required for the sides and back of the collector. Insulating materials must be durable, heat resistant, and thermally efficient. Thermal efficiency is specified by either the *K factor*, which measures the thermal *conductivity* of the material, or an *R value*, which is the reciprocal or 1/K. The *lower* the K factor (higher the R value), the better an insulator the material is.

Table 2-9 provides a K factor comparison for different materials. These values represent 1-inch thick material and must be divided by the overall thickness (in inches) specified, to arrive at a total value. For example, I call for 2 1/2-inch thick fiberglass on the back of the absorber. From Table 2-9, which provides figures for one inch thickness, you can see that this material in a 2 1/2-inch thickness would have a K factor of 0.1 (0.24/2.5) and an R value of 10 (1/0.1).

Regular fiberglass building insulation used for walls, ceilings, etc. has a Bakelite binder which will smell and smoke if heated above 250° F. Because the absorber of a solar collector may get as hot as 400° F in the sun when no water is flowing, I have recommended using an industrial grade of fiberglass board made for higher temperatures. Such a product is Johns-Manville No. 814 or CertainTeed Corp. No. 850. These materials are available in a 24″ × 48″ standard size from building contractors and they cost approximately $0.21 per square foot in a 1 inch thickness. You should *not* buy the material faced with foil and paper called FSK (foil scrim-Kraft) or FRK (foil-reinforced-Kraft).

Closed-cell foamed rubber is suitable for insulating piping between the collectors and storage tank. As you can see from Table 2-9, this material is not as good an insulator as urethane, fiberglass, or polystyrene, but in 1-inch thickness it does have an R value of 3.6 which should be suitable in most climates. This material is flexible, moisture resistant, and easily fabricated and sealed with adhesive. One manufacturer, Rubatex Corp., provides standard inside diameters from 3/8 inch to 5 1/2 inches, with wall thicknesses of 1/2 inch and 3/4 inch. For greater wall thickness, two layers will be required. Six-foot lengths are generally available from stock at heating and ventilating dealers and contractors.

PAINT

As will be detailed in Chapter 6, you have two choices on selection of a flat-black paint for the absorber: either a heat-resistant flat-black designed for painting barbecues or automobile engines, or the industrial grades designed for solar collector applications. Even though they are more expensive and require a primer, I suggest the products made by DeSoto, Inc. called Enersorb or the 3M Company called Nextel. The Enersorb product comes in both quart and gallon sizes. A quart kit contains 1 quart of black base and 1 quart of clear activator mixed together in equal volume just before use. The cost of the kit is $20 FOB factory Des Plaines, Ill.

HEAT TRANSFER FLUID

Water is the best heat transfer fluid in many respects: it is inexpensive, nonflammable, nonpoisonous, and has a high heat capacity. It has one major disadvantage—it freezes at 32° F. For the collector designs in Chapter 6, 7, and 9 water has been specified. For the Chapter 8 system, a buffered and inhibited 50/50 solution of ethylene glycol and distilled water is acceptable if a dual-walled heat exchanger is used and the fluid is periodically drained and replaced. This mixture will provide freeze protection to approximately −34° F.

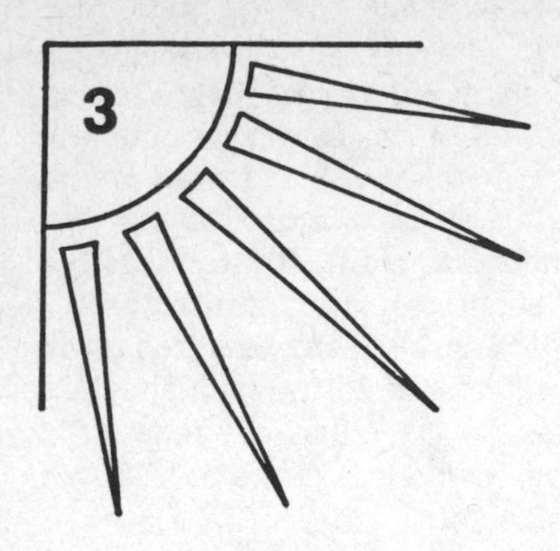

Description and Availability of Assembled Components

This chapter will include typical manufacturers' specifications on a few quality components. Skills, material costs, your decision on whether to build or buy and other factors to consider for solar collectors will all be covered in other chapters.

There are many good brands that are not included in these pages, but it is not possible to show them all. For further considerations, a reference catalog to solar products previously published by ERDA (Energy Research and Development Administration) has been updated and is now available from the Solar Energy Industries Association (SEIA), 1001 Connecticut Avenue, N.W., Washington, D.C. 20036. This 381-page comprehensive guide to manufacturers and service organizations is entitled *Solar Industry Index*, and it was priced at $8 in March of 1977.

There is another good reference document available: the *Florida Solar Energy Equipment and Services Directory*, which lists equipment, manufacturers, distributors, retailers, and service organizations for solar energy. The Florida Solar Energy Center, which publishes this booklet, was chartered by the State Legislature in 1974 to disseminate information on solar products. As in the case of the SEIA catalog, a strong effort was made to authenticate the information provided, and the result is a comprehensive reference base which describes the extent, activity, and technological capability of Florida's solar energy industry. This 100-page directory can be obtained by writing to the Florida Solar Energy Center, 300 State Road 401, Cape Canaveral, Florida 32920.

Component pricing varies widely between distributors as a function of shipping costs, quality, etc. My suggestion would be to select those items in which you have an interest and check with the manufacturers' authorized dealers or distributors locally for price and availability. A comparison on solar collectors can then be made on the advantages of a home-built and commercially available units.

The following pages contain manufacturers' specifications for some quality components.

PPG INDUSTRIES

General Description

PPG Standard Solar Collectors are available with:

(1) One (1) or two (2) clear 1/8-inch glass cover plates. These cover plates are heat treated for improved breakage resistance.
(2) A coated metal absorber plate with fluid carrying passages.
(3) Back insulation and a protective box (optional).
(4) Unit size: 34 3/16″ × 76 3/16″ × 1 3/8″ (without optional back insulation and box).

High quality PPG Standard Collectors are factory fabricated and sealed to preclude internal condensation. Figure 3-1 illustrates a typical edge section of a two-glass cover plate PPG Standard Solar Collector.

Properties of PPG Standard Solar Collectors

1. *Glass Cover Plates:* 1/8-inch clear heat-treated glass is used for PPG Standard Solar Collectors. The properties of 1/8-inch clear, heat-treated glass are as follows:
   (a) Total Solar Energy Transmittance: 85% (single 1/8-inch thick).
   (b) Total Solar Energy Reflectance: 8 % (single 1/8-inch thick).
   (c) Hemispherical Emissivity, $\epsilon_h = 0.84$ (0 – 200F).
   (d) Specific Heat, $C_p = .205$ (32–212F).

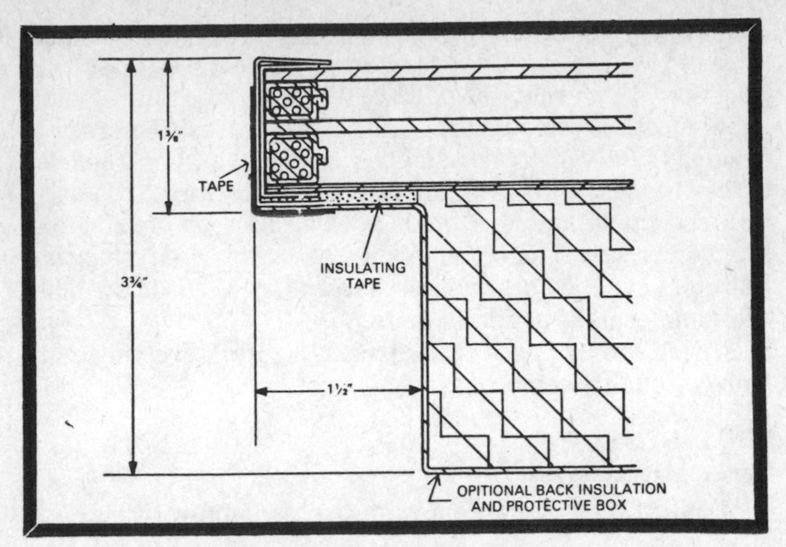

Fig. 3-1. Typical edge section of PPG standard solar collector.

(e) Wind Load: The 34 3/16 × 76 3/16 insulating unit (2 glass cover plates) for the PPG Standard Solar Collector can withstand a one-minute, fastest mile uniform wind load of 102 psf with a probability of breakage equal to 8 lights per 1000.

(f) Snow Load: The 34 3/16 × 76 3/16 insulating unit (2 glass cover plates) for the PPG Standard Solar Collector can withstand a long-term (more than one hour) uniform snow load of 68 psf, with a probability of breakage equal to 8 lights per 1000.

(g) Hail: PPG 1/8-inch Herculite glass has been shown to be effective against hailstones.

2. *Metal Absorbing Plate:* Metal absorbing plates in PPG Standard Solar Collectors are coated with a flat-black coating, absorptivity of 0.95, and hemispherical emissivity of 0.88. This coating is PPG's Duracron 600 L/G, which is sold exclusively to and applied by licensed applicators. Selective coatings for PPG Standard Solar Collectors are available on request.

PPG metal absorbing plates are available in both aluminum and copper. Please specify metal type when ordering PPG collectors.

3. *Insulation and Protective Metal Pan* (Optional): If the responsible professional desires back insulation, PPG Standard Solar Collectors can be factory fabricated with fiberglass insulation housed in a protective metal pan. The fiberglass insulation is 2 1/2 inches thick, 3 lb. density (thermal conductivity, k = 0.24 Btus per square foot per inch of thickness per degree Fahrenheit). The protective metal pan is required to protect and hold the fiberglass insulation in place.
4. *Edge Configuration:* The sealants, spacers, desiccant, and metal channel used to fabricate PPG Standard Solar Collectors have been especially selected and fabricated to reduce edge degradation, pressure buildup, and internal moisture condensation.

Table 3-1 compares the calculated collector parameter with the efficiency for PPG Standard Solar Collectors, and Fig. 3-2 shows the PPG Standard Solar Collector's instantaneous efficiency at an air temperature of 50°F.

PPG Industries, Inc. Solar Systems Sales are located at One Gateway Center, Pittsburgh, Pennsylvania 15222.

SUNWORKS

Following are technical data on Solector Solar Energy Collectors, both liquid-cooled and air-cooled.

Table 3-1: Calculated Collector Parameter vs. Efficiency for PPG Standard Solar Collectors (45° tilt, south facing, wind = 7.5 mph).

COLLECTOR PARAMETER (°F-hr-SQ. FT./BTU)	INSTANTANEOUS EFFICIENCY (%)
.05	68
.10	65
.15	62
.20	58
.25	55
.30	50
.35	46
.40	43
.45	36
.50	32
.55	26
.60	20
.65	14
.70	7

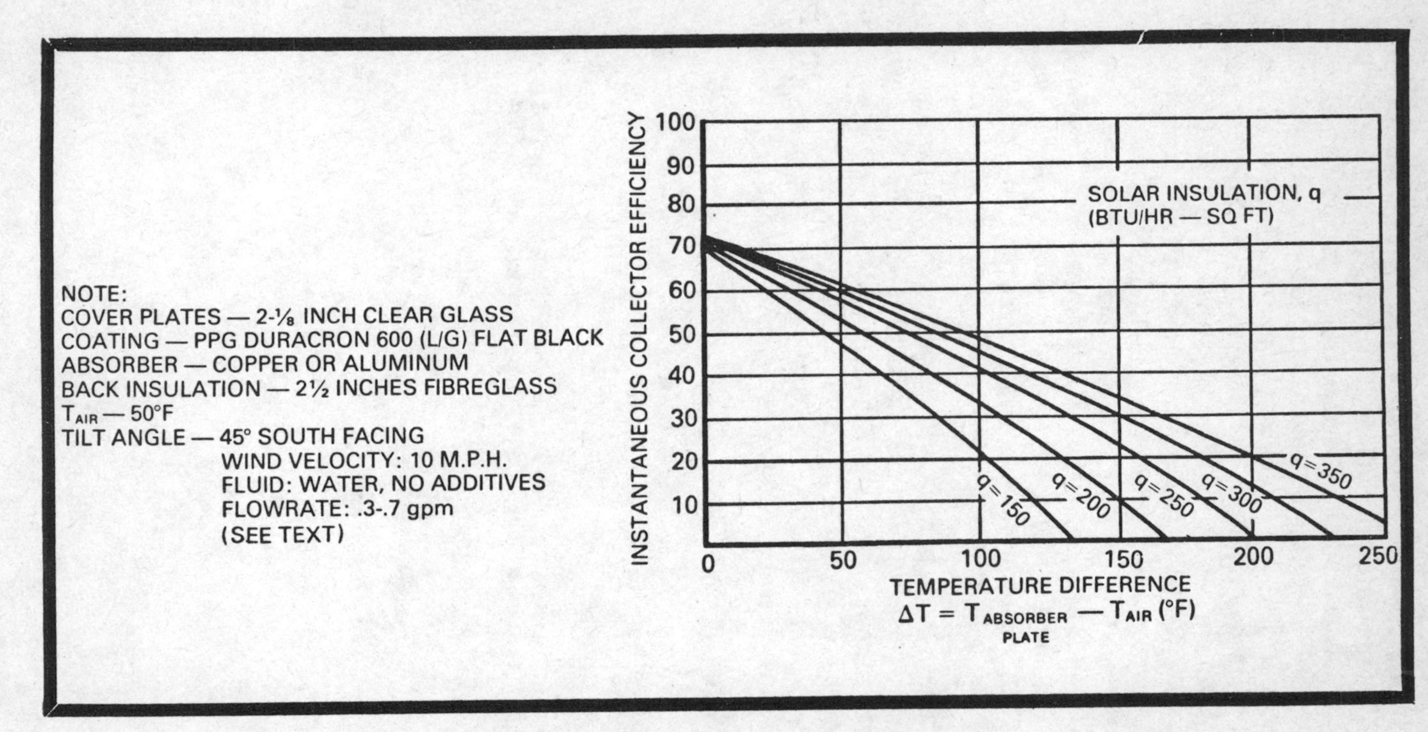

Fig. 3-2. PPG standard solar collector instantaneous efficiency, for T_{air}=50°F.

Technical Data

Cover: Single glazing, no iron content, 3/16 inch tempered, edges swiped; double glazing, no iron content, 2 1/8 inch tempered, edges swiped, sealed unit. Total solar transmissivity of single glazing = 92 percent, double glazing = 85 percent (Fig. 3-3).

Absorber container: Sides, aluminum extrusion; rear, aluminum sheet 0.05 inch thickness; pop rivet in place.

Air space between cover and absorber: Approximately 3/4 to 1 inch.

Gasketing material: Neoprene "U" gasket.

Weatherproofing: This module can be placed out in the weather without need for further weatherproofing.

Finish on aluminum sides of container: Standard mill finish. Anodized clear or black finish available at extra cost.

Dimensions of surface-mounted module: Overall outside dimensions are 35 1/2 inches wide by 84 inches long by 4 inches thick (add 1 1/4 inches at each end for continuous mounting bracket liquid Solector only). Effective absorber area is 18.68 square feet. Ratio of usable absorber area to total surface covered is 0.902. Glass area is 18.96 square feet.

Solector solar energy collectors can be mounted end-to-end for series flow or side-by-side for parallel flow. It is recommended that no more than three collectors be connected in series. The Solector solar energy collector modules for both liquid and air are identical in size: 3 feet wide and 4 inches thick. They are available in two lengths, 5′4″ long or 7′0″ long.

Liquid Data

Absorber: Copper sheet is 0.010 inches thick (7 ounces); selective black has a minimum absorptivity of .87/.92 and a maximum emissivity of .07/.35, manufactured by Enthone Inc., guaranteed durable to 400° F; copper tubes are 1/4 inch I.D. (0.375 inches O.D.), L-type; tube spacing is 6 inches on center; tube pattern is a grid; bond between the tube and sheet is high temperature solder; manifolds are 1 inch I.D. (1.125 inch O.D.), M-type copper; tube connections to manifold are brazing alloy; and connection to external piping is 1 inch I.D. (1.125 inch O.D.) K-type copper, extending 1 7/8 inch beyond collector ends. Supply is top right, return is bottom left (Fig. 3-4).

Insulation behind absorber: 2 1/2 inch thick fiberglass, 1.5 pound per cubic foot density; R = 10.0.

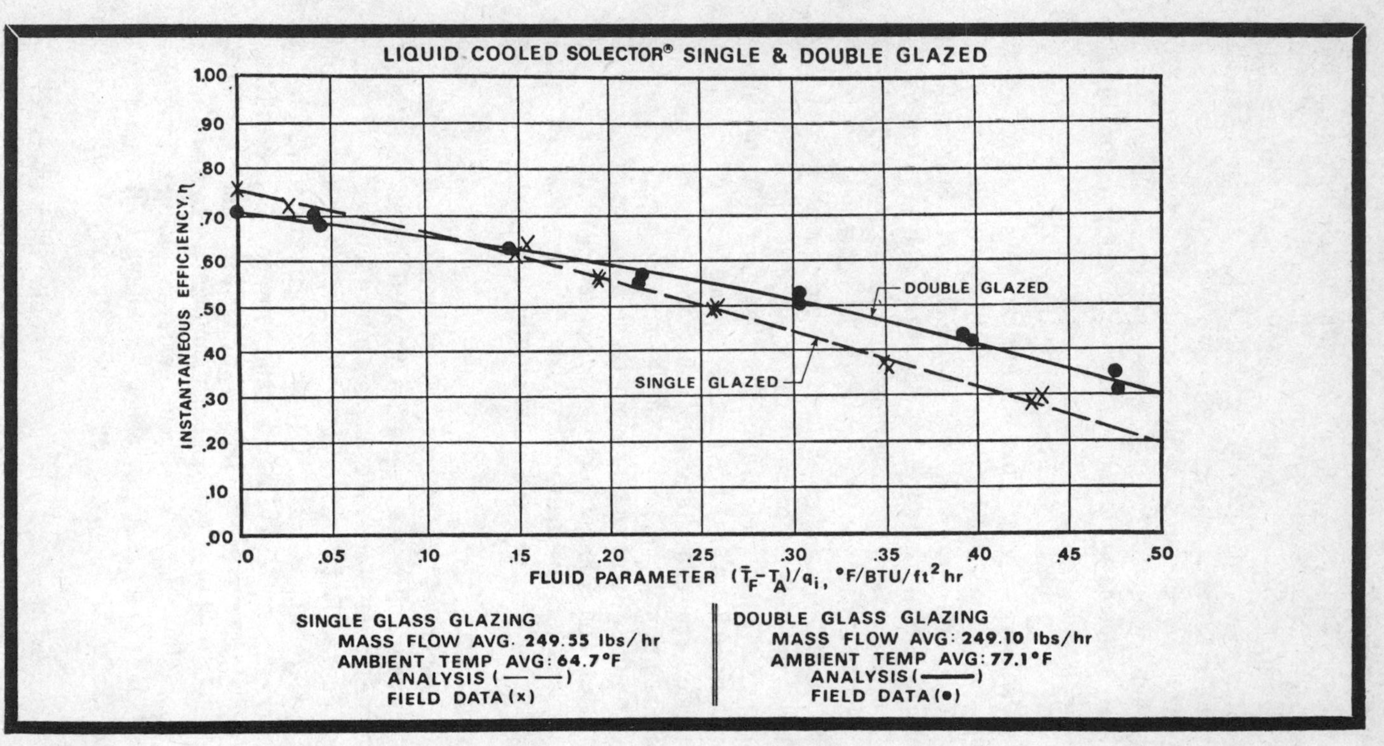

Fig. 3-3. Liquid-cooled Solector, single and double glazed.

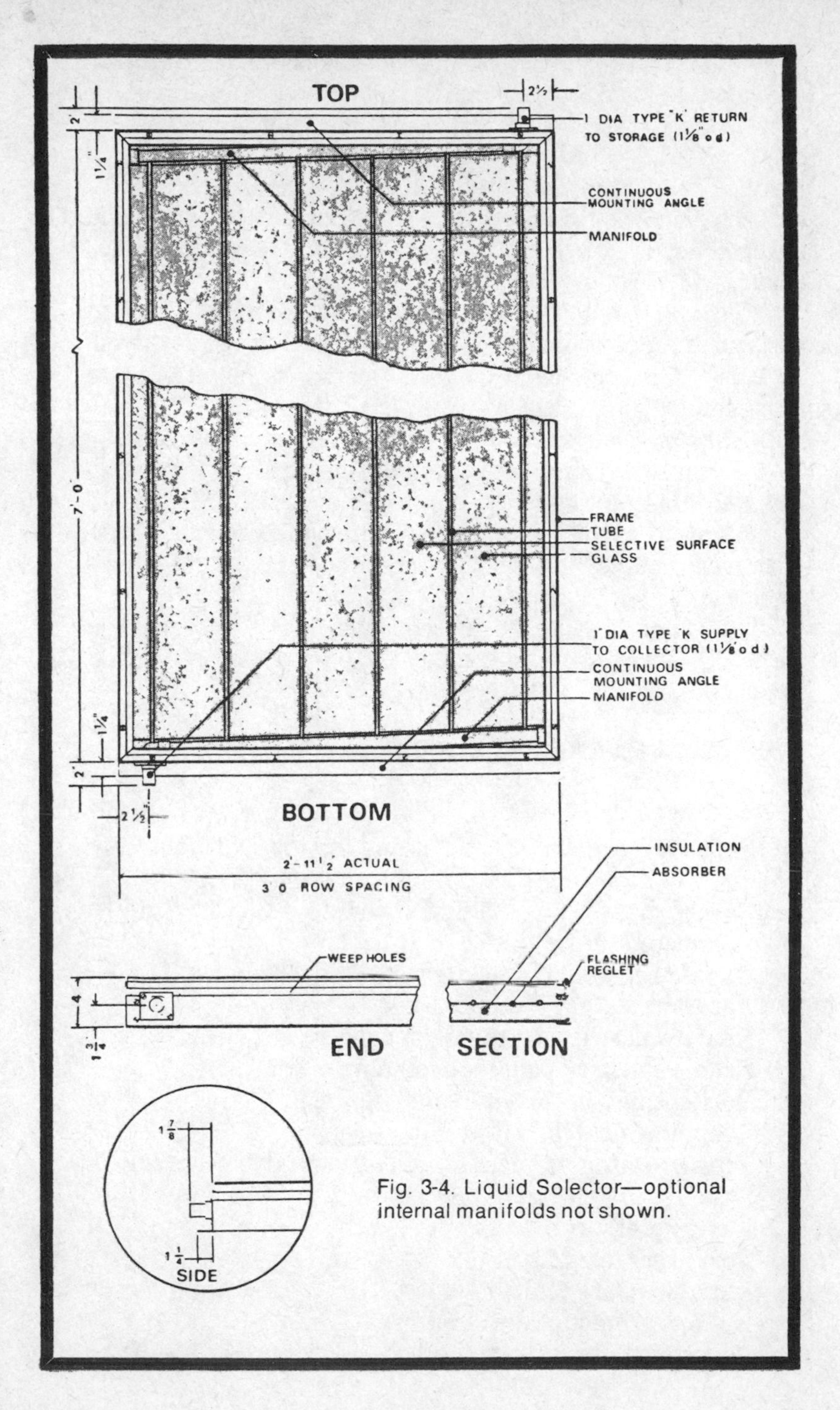

Fig. 3-4. Liquid Solector—optional internal manifolds not shown.

Method of anchoring: Continuous mounting bracket at each end of frame for anchoring; four predrilled holes are provided for anchor bolt or screw connections; additional holes may be drilled by installer if required.

Weight per module: 113.9 pounds, filled; 111 pounds, empty. Add 27 pounds for double-glazed unit (Note: the liquid in the collector is equal to 0.36 gallons or 46.4 ounces or 2.90 pounds or 0.05 cubic feet or 80.5 cubic inches.)

Recommended flow rate through collector: 14.7 pounds per square foot per hour (.55 gpm) per collector. Single glazed, $F_r = .90$; double glazed, $F_r = .93$. Flow resistance at this rate is negligible.

Collector coolant: Coolant should be Sunsol 60 made by Sunworks. In areas where regular tap water is used as a coolant, it is important that the pH be controlled between 6.5 and 8. These collectors can be used with other coolants but the user must contact the manufacturer for approval of specific liquids.

A division of Enthone Incorporated, and a subsidiary of ASARCO, Sunworks is located at P.O. Box 1004, New Haven, Connecticut 06508.

LOF SOLAR ENERGY SYSTEMS

Specifications

Following are the specifications for the LOF unit (Fig. 3-5):

Cover glass (2), fully tempered, 1/8-inch Tuf-Flex tempered safety glass.
Exposed absorber plate area: 19.47 square feet.
Absorber plate: all copper.
Heat transfer fluid system: all copper.
Frame and back panel: aluminum.
Back insulation: 3 inch treated fiberglass.
Recommended fluid flow rate: 0.5 gpm.
Pressure drop at recommended flow (glycol/water): 0.5 psi.
Design pressure: 100 psi.
Text pressure: 200 psi.
Burst pressure (calculated): 450 psi.
No-flow protection required: none.
Maximum internal temperature (no-flow): 450° F.

Performance test specification: NBSIR 74-635.
Environmental conditions: −40° F to 120° F.
Maximum wind or snow load: 30 pounds per square foot.
Fluid connections: brass 45° flared.
Installation orientation: short side horizontal.
Minimum tilt angle to allow complete draining of panel: 5° from horizontal.
Maximum tilt: 90°.
Mounting: 4-point or brackets.
Panel weight (filled): 150 pounds.
Installed load: 7.5 pounds per square foot.
Shipping weight: 160 pounds.

Warranty

These units come with the following warranty: "We warrant that our SunPanel solar collector panel will meet our published specifications and will be free of defects in the material and workmanship for a period of one year from the date of manufacture. This warranty shall not apply to solar collectors which have not been handled, installed, or used in

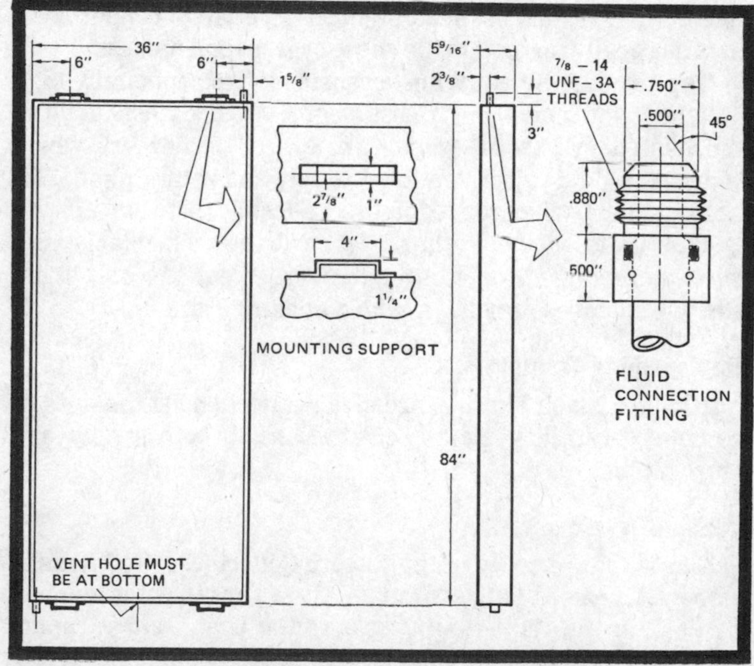

Fig. 3-5. SunPanel, an LOF solar energy system.

accordance with our published instructions, or to the replacement parts or panels beyond the warranty period applicable to the original solar panels. Our liability under this warranty shall be limited to replacement of the material, FOB, the shipping point nearest the installation, or at our option to refund the purchase price, and we shall not be liable for cost of removal or installation of or any consequential damages or other damages in connection with such collector panel. We make no warranty of merchantability, no warranty that SunPanel is fit for any particular purpose or use and no other warranty is expressed or implied."

The Libbey-Owens-Ford Company is located at 811 Madison Avenue, Toledo, Ohio 43695.

REVERE MODULAR SOLAR COLLECTORS

Revere Modular Solar Energy Collectors are simple, competitively priced, and they incorporate a number of good features in their construction.

Tube-in-Strip Collector Plate

The collector plate is constructed of Revere's patented Tube-in-Strip which is a solid sheet of quality copper with tubes integrally formed in the sheet by a special process.

The result is a heat transfer surface of exceptionally high efficiency and corrosion resistance. The tubes conduct the fluid heated by the sun. They are brazed to headers which conduct the fluid to conventional supply and return piping. To improve the performance of the collector plate on all our standard units, they coat the surface with a special black paint which increases absorption of the sun's heat. As an option, selective surfaces may be provided at extra cost.

Simple Piping Connections

Flush 1/2-inch female threaded connections at the ends of the collectors allow simple connections to both supply and return piping.

Transparent Cover

The standard cover of the Revere Collector is two layers of tempered glass which permits the sun's rays to enter and heat the fluid in the tubes, but prevents the heat from escaping. Single layer covers are also available for use determined by geographical location and unit efficienty requirements.

Table 3-2. Average Collector Efficiencies and Outputs for Various Applications.

APPLICATION	SOLAR INPUT	SEASON	AVERAGE EFF. (%)	AVG. OUTPUT—BTU/HR./SQ. FT.
Swimming Pools	Very Good Good	Summer	70 60	175 120
Heating Domestic Water	Very Good Good	Non-Summer	50 45	125 90
	Very Good Good	Summer	60 50	150 100
Heating	Very Good Good	Spring/Fall	45 40	115 80
	Very Good Good	Winter	35 30	90 60

Solar Input: Very Good—250 BTU/Hr./Sq. Ft.
Good—200 BTU/Hr./Sq. Ft.

Insulated Housing

Revere's copper collector is well insulated and completely enclosed in a sturdy Revere extruded aluminum housing.

Specifications

Overall dimensions: 34 7/8″ × 76 7/8″.
Net effective area of collector plate: 17.2 square feet.
Weight with double glass: 115 pounds.
Recommended flow rate: 0.5 to 1 gpm.
Unit connectors: 1/2″ nom. F.T.P.
Thickness of copper collector plate: .032 inch.
Tube spacings: 5 1/2 inches on centers.

Table 3-2 summarized average collector efficiencies and outputs for various applications, and Fig. 3-6 illustrates the Revere Sunaid solar collector.

Revere Copper and Brass Incorporated's solar energy department can be reached at P.O. Box 151, Rome, New York 13440.

RHO SIGMA INCORPORATED

RS 500 Model

The proportional solar control of this unit has an input of 120VAC and an output of 120VAC, 6 amps. The RS 500P-1HL model has all solid state electronic circuitry which varies speed as a function of the ΔT to achieve maximum efficiency in energy transfer. Minimum flow: $\Delta T = 3° \pm 1°$ F. Full flow: $\Delta T = 12° \pm 1°$ F.

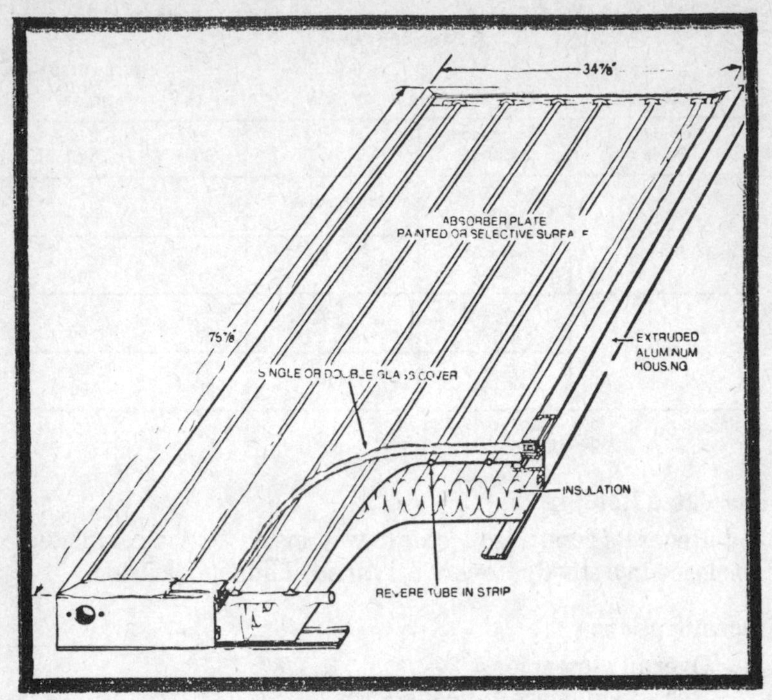

Fig. 3-6. Revere Sunaid solar collector.

Power delivered to the pump is at full voltage and zero crossover, thus assuring full torque at low speed and the absence of line noise. The design eliminates motor speed hysterisis and assures smoothly variable motor speed control. Compatible with permanent-capacitor and shaded-pole motors.

Standard features include a pulsing indicator light to indicate pump speed; a switch which meets local electrical code requirements for a pump power disconnect switch when the control is located within 6 feet of the pump; a high temperature turn-off to limit the upper temperature of the storage tank (140° F is standard); and low temperature turn-on circuits to turn on the pump at a specified low temperature to protect solar collectors from freeze damage (37° F is standard).

The RS 500P-1H/L-21/H control has two 6-amp outputs. The first output provides proportional control with optional high and low temperature override circuits identical to the RS 500P-1HL. The second output is normally configured to break the 110VAC power to drain valves on the approach of freezing conditions. But the second output can also be configured to customer specifications to control a second pump or numerous other valve configurations including 24VAC power. Additional flexibility is provided through various interlocks between the two outputs. Other options include power delivery through the second output to activate various cooling mechanisms on over-temperature signals from either the collector of the storage sensors.

Sensors

All Rho Sigma sensors are electrically identical and interchangeable and are designed to withstand stagnation temperatures of solar collectors. Only two sensors are required with each differential thermostat.

The *SA Sensor* is the temperature sensing element encased in epoxy (Fig. 3-7).

The *ST Sensor* has a copper housing with a hole punched in it for bolting directly to the collector plate or suspending inside air ducts. Alternatively, a radiator hose pipe clamp may be used to secure the rugged sensor to the surface of a pipe. Or it may be slipped inside the insulation of the storage tank.

The *SF Sensor* is a 1″ × 1″ × 2″, sandblasted and black-anodized aluminum sensor designed for use with unglazed solar collectors or in high flow rate, low delta-T systems. The screw provided with the sensor may be used to mount the sensor near the collector where it will sense the temperature and availability of solar energy at the collector. Designed primarily for all use with the RS 240.

The *SPT-XX Sensor* has a probe at the end of its 1/2″ pipe threads. The temperature-sensing element is at the tip of the all-brass sensor. It is designed primarily for insertion into tanks to obtain the most accurate measurement of the fluid or air temperature inside. Standard probe lengths are 1 1/2″, 3″, 4 1/2″, 6″, 12″, and 24″.

The *SP Sensor* is epoxied into a rugged brass housing with standard 1/2 inch pipe threads for easy installation into standard plumbing fixtures.

The *SPR Sensor* may be screwed into the end of a pipe which may be inserted into the top of a deep tank. Wires run inside the pipe to the control. Provides accurate sensing of temperature at the bottom of deep tanks.

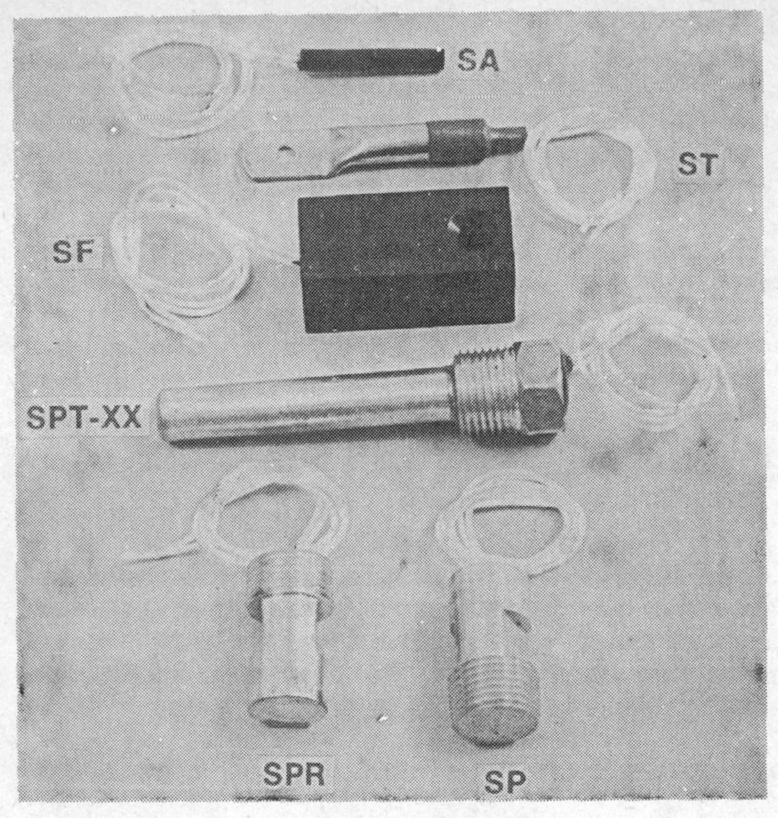

Fig. 3-7. Rho Sigma sensors.

Dual Thermistors may be specified in any housings. One thermistor provides control signals; the other thermistor may provide signals for instrumentation or a second solar control (e.g., Model SP-2). Sensors will provide strong signals even when separated from the controller by 200 feet.

Rho Sigma, Inc. is located at 15150 Raymer St., Van Nuys, CA 91405.

GRUNDFOS

Stainless Steel Circulator Pump—UP 25-42 SF

The UP 25-42 SF is a revolutionary circulator pump. The water passing through the pump touches nothing but high quality fabricated stainless steel. The volute section, for example, is constructed of type 316 stainless. As with all Grundfos circulators, the UP 25-42 SF is engineered to be interchangeable with the pumps of all other major manufacturers.

Construction: The UP 25-42 SF is a water lubricated pump. However, in order to protect the rotor and bearings from damaging impurities which may be present in the circulating water, they are separated from the stator and the pump chamber by a liquid-filled rotor can. The motor shaft extends out from the rotor can, into the pump chamber through the aluminum oxide bearing, which also functions as a seal. During initial operation, the pump is automatically self-vented; however, due to the isostatic principle, there is no further recirculation of water into the closed rotor can.

The pump's "diamond-hard" aluminum oxide bearing construction, combined with the high starting torque of the motor, insures re-start after shutdown.

Materials: The pump contains the following materials:

Stainless steel: Pump chamber, rotor can, shaft, rotor cladding, bearing plate, impeller, thrust bearing cover.

Aluminum oxide: Top bearings, shaft ends, bottom bearing.

Carbon/aluminum oxide: Thrust bearing

Aluminum: Motor housing, pump housing cover.

Ethylene/propylene rubber: O-rings, gasket.

Silicone rubber: Winding protection.

Applications: The UP 25-42 SF is particularly suited for open and potable systems. The stainless steel construction protects the pump from the corrosion that has plagued cast iron and bronze-lined pumps in these types of applications. The pump is intended for circulation and booster applications in domestic water systems.

PERFORMANCE CURVE FOR UP 25-42 SF

Electrical and Performance Data: The UP 25-42 SF is operated by an energy-conserving 1/20th hp (0.85 amp) motor which has built-in overload protection. However, because of advanced engineering design, the pump produces up to 14 feet of head or a flow of up to 23 gpm. The pump's small size and high efficiency make it suitable for many applications, and greatly reduces installation problems (Fig. 3-8).

Dimensions of the UP 25-42 SF are shown in Fig. 3-9.

Isolation Valves: Grundfos recommends the use of isolation valves with circulation pumps in all systems (Fig. 3-10).

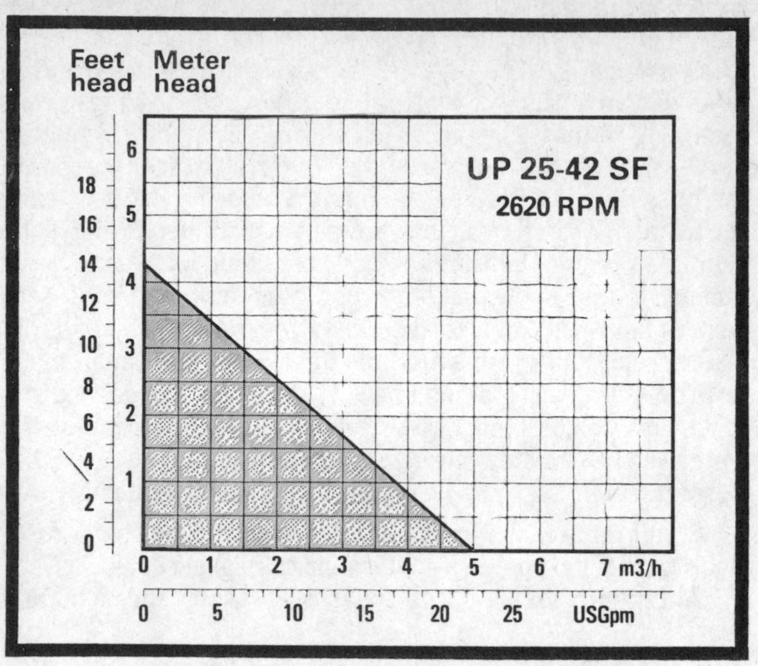

Fig. 3-8. Performance curve for the Grundfos UP 25-42 SF.

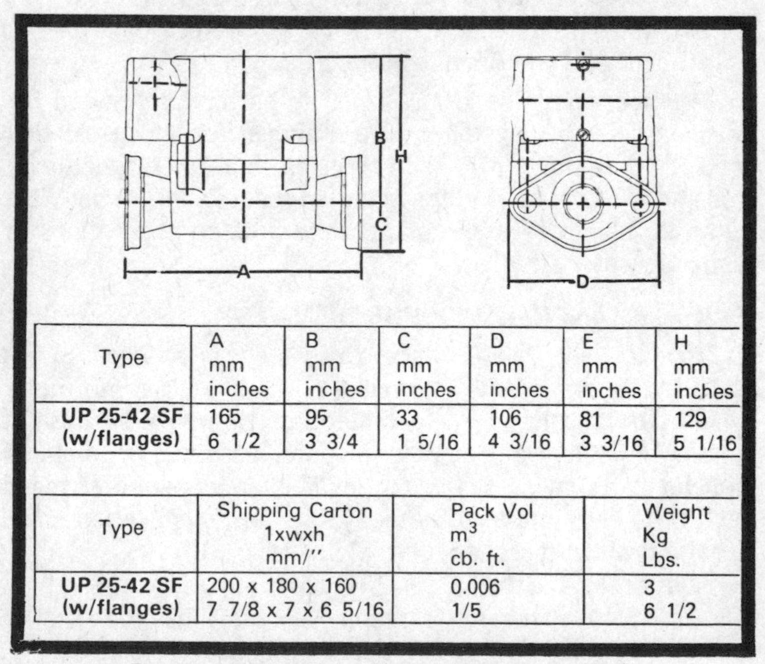

Type	A mm inches	B mm inches	C mm inches	D mm inches	E mm inches	H mm inches
UP 25-42 SF (w/flanges)	165 6 1/2	95 3 3/4	33 1 5/16	106 4 3/16	81 3 3/16	129 5 1/16

Type	Shipping Carton 1xwxh mm/"	Pack Vol m^3 cb. ft.	Weight Kg Lbs.
UP 25-42 SF (w/flanges)	200 x 180 x 160 7 7/8 x 7 x 6 5/16	0.006 1/5	3 6 1/2

Fig. 3-9. Dimensions of the Grundfos UP 25-42 SF.

Two-Speed Circulator Pump—UPS 20-42

The UPS 20-42 is fitted with a variable flow control and also features a two-speed motor (Fig. 3-11). The head is controlled by the flow adjustment arm (I) and the choice of speed is made by hand on the switch (B) or made automatically in conjunction with remote control.

Construction: The UPS 20-42 is a water lubricated pump. However, in order to protect the rotor (P) and bearings (E,L) from damaging impurities which may be present in the circulating water, they are separated from the stator (F) and the pump chamber by a liquid filled rotor can (D). The shaft (Q) extends out from the rotor can, into the pump chamber through a shaft seal which also functions as a thrust bearing (M). During initial operation the pump is automatically self-vented; however, due to the isostatic principle, there is no further recirculation of water into the closed rotor can.

The bearing construction combined with the high starting torque of the motor, insures re-start after shutdown.

Materials: The pump contains the following materials:

Stainless steel: Rotor can, shaft, rotor cladding, bearing plate, impeller, variable flow adjustment plate, thrust bearing cover.

Aluminum oxide: Top bearing, shaft ends, bottom bearing.

Carbon/aluminum oxide: Thrust bearing.

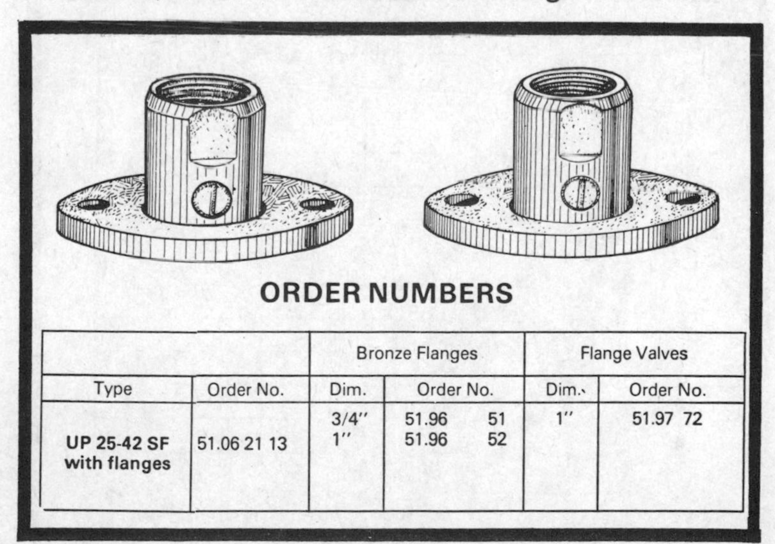

ORDER NUMBERS

		Bronze Flanges		Flange Valves	
Type	Order No.	Dim.	Order No.	Dim.	Order No.
UP 25-42 SF with flanges	51.06 21 13	3/4″ 1″	51.96 51 51.96 52	1″	51.97 72

Fig. 3-10. Flange isolation valves.

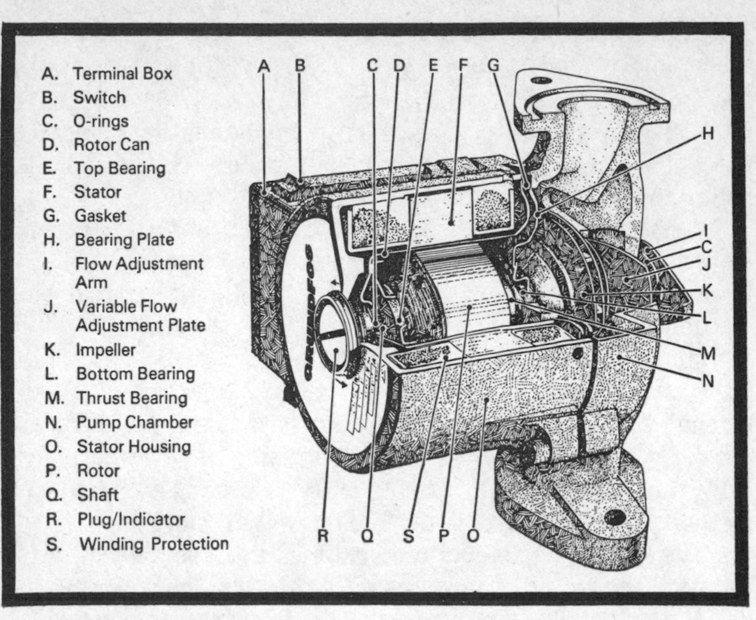

Fig. 3-11. The Grundfos two-speed circulator pump, UPS 20-42.

Cast iron: Pump housing.
Ethylene/propylene rubber: O-rings, gasket.
Silicone rubber: Winding protection.

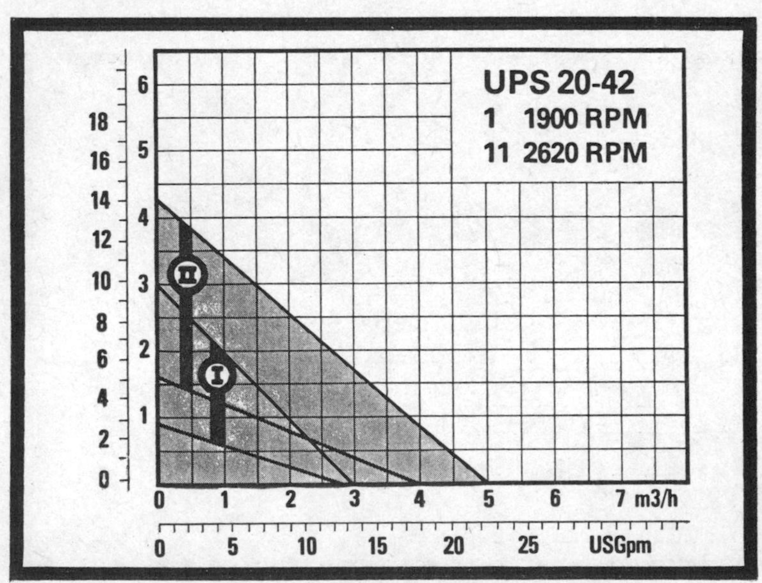

Fig. 3-12. Performance curves for the Grundfos UPS 20-42.

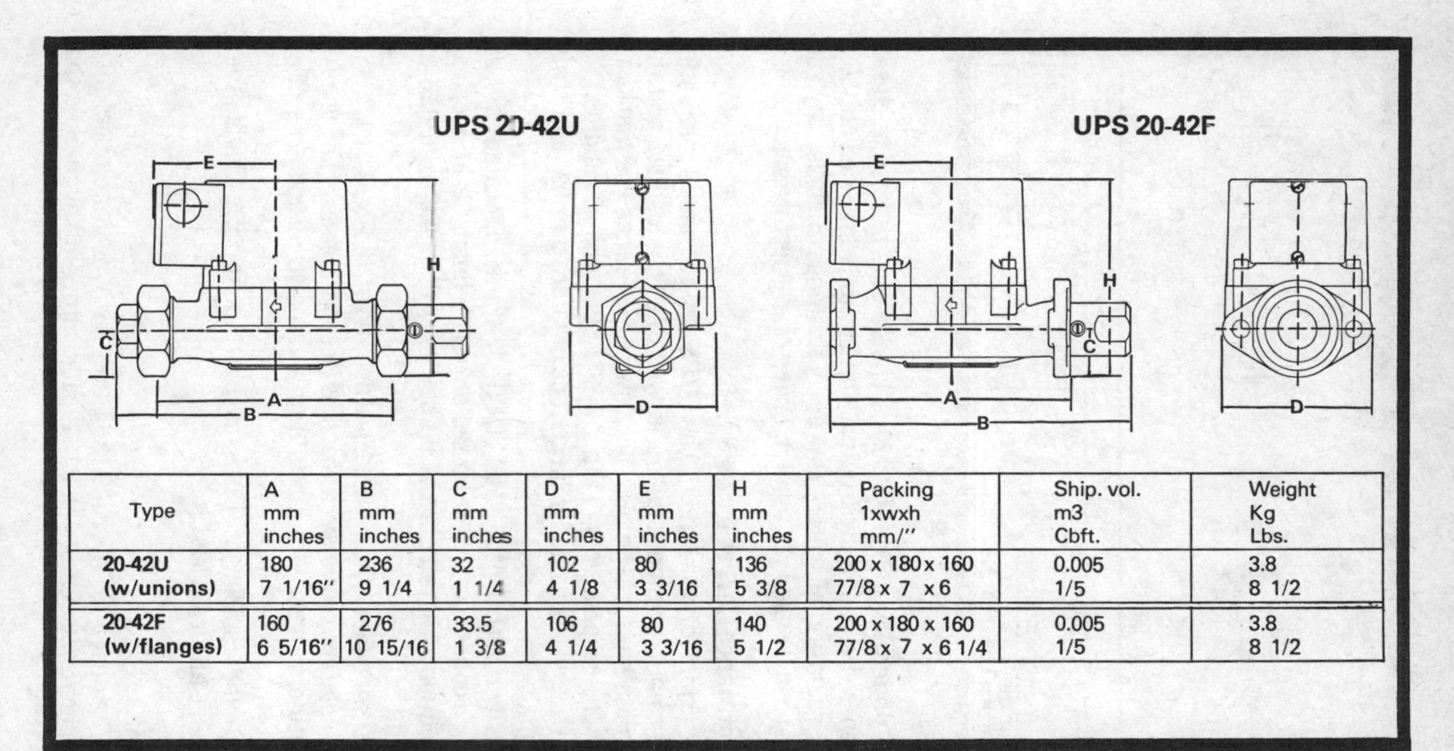

Type	A mm inches	B mm inches	C mm inches	D mm inches	E mm inches	H mm inches	Packing 1xwxh mm/″	Ship. vol. m3 Cbft.	Weight Kg Lbs.
20-42U (w/unions)	180 7 1/16″	236 9 1/4	32 1 1/4	102 4 1/8	80 3 3/16	136 5 3/8	200 x 180 x 160 77/8 x 7 x 6	0.005 1/5	3.8 8 1/2
20-42F (w/flanges)	160 6 5/16″	276 10 15/16	33.5 1 3/8	106 4 1/4	80 3 3/16	140 5 1/2	200 x 180 x 160 77/8 x 7 x 6 1/4	0.005 1/5	3.8 8 1/2

Fig. 3-13. Dimensions for the Grundfos UPS 20-42.

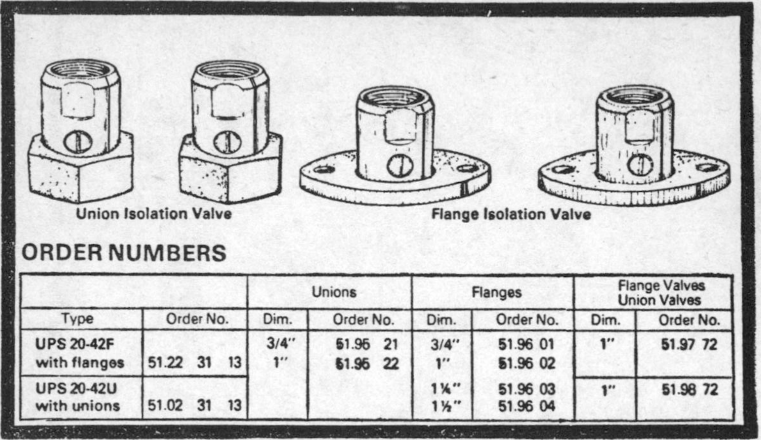

ORDER NUMBERS

		Unions		Flanges		Flange Valves Union Valves	
Type	Order No.	Dim.	Order No.	Dim.	Order No.	Dim.	Order No.
UPS 20-42F with flanges	51.22 31 13	3/4" 1"	51.95 21 51.95 22	3/4" 1"	51.96 01 51.96 02	1"	51.97 72
UPS 20-42U with unions	51.02 31 13			1¼" 1½"	51.96 03 51.96 04	1"	51.98 72

Fig. 3-14. Isolation valves.

Applications: The UPS 20-42 should only be used in closed systems (i.e. solar, hydronic). The pump is intended only for the circulation of water. However, solutions such as ethylene glycol can be used without hindering pump performance.

Performance Curves for UPS 20-42

Performance Data: The UPS 20-42 has a versatile performance range due to the variable flow control and the dual rpm switch. The high and low rpm settings are marked I and II respectively (Fig. 3-12). Contact Grundfos for information regarding larger circulator pumps and twin pumps.

Electrical Data: The UPS 20-42 is operated by an energy-conserving 1/20 hp motor which has built-in overload protection. The amperage on setting "I" is 0.65 and 0.85 on setting "II".

Dimensions of the UPS 20-40 are shown in Fig. 3-13.

Isolation Valves: Grundfos recommends the use of isolation valves with circulation pumps in all systems (Fig. 3-14).

Grundfos Pumps Corp. is located at 2555 Clovis Ave., Clovis, California 93612.

TACO

The Taco SSM (Figs. 3-15 and 3-16), is a compact, prepackaged, pumping, heat transfer and control module that features:

- Built-in pump controls
- Hi-efficiency Solarchanger heat exchanger
- Glycol and water pumps
- Can be wall, floor, or tank mounted
- Needs only one electrical connection
- Requires only four simple piping connections
- Compact

The unit replaces lengthy, costly field assembly of:

- Two pumps
- An expensive tank unit
- Controls
- Expansion tank
- Piping

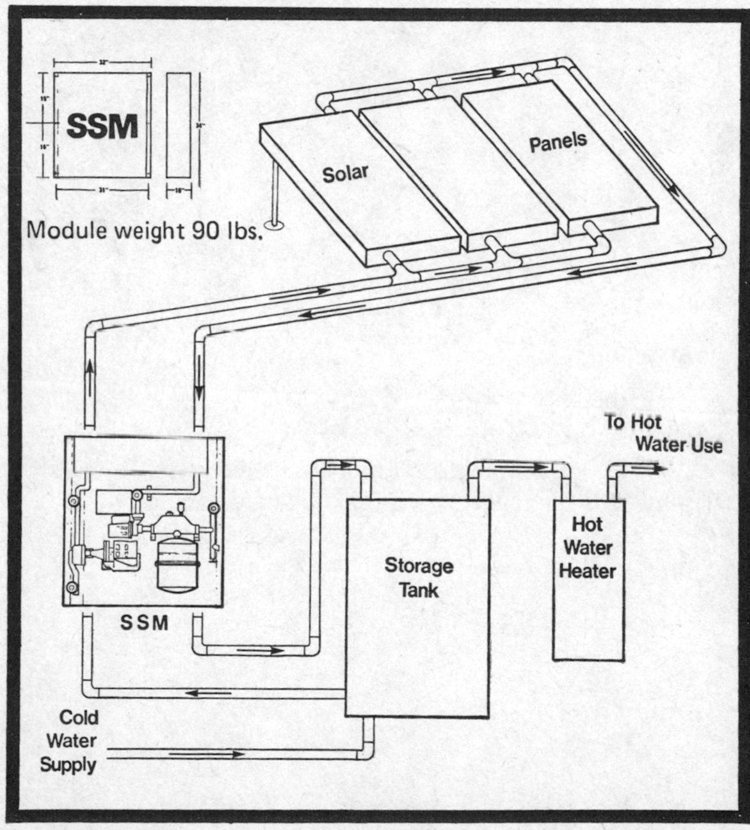

Fig. 3-15. The basic solar-assisted domestic hot water system utilizing Taco SSM solar systemizer.

Fig. 3-16. The Taco SSM.

- Wiring
- Brackets

Taco Incorporated is located at 1160 Cranston St., Cranston, Rhode Island 02920, and Taco Heaters of Canada, at 3090 Lenworth Drive, Mississauga, Ontario, Canada.

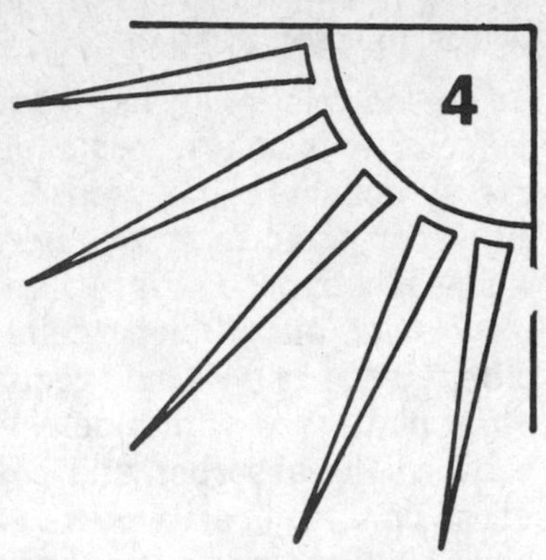

Solar Collectors: Build or Buy?

In addition to assessing whether you have the skill and tools to construct your own collectors, there are several other factors to consider before deciding to build rather than buy the units. This chapter will highlight them for you.

MATERIAL COST

Aside from the satisfaction of making something with your own hands, do-it-yourself projects should save money on labor. Your materials, however, will cost *more* than that paid by the collector manufacturer.

First you should decide on the collector design that is best suited for your climate and system. (Details on this issue appear in subsequent chapters). Then, using the *raw material cost* listed in Chapter 2 as a guideline, develop the cost per square foot of collector from that standpoint. Next, confirm the price and availability of the materials needed from your own supply sources, because these will vary considerably.

You should also consider a *scrap factor* for remnants, overages, and poor workmanship. The drop-off, or remnants from cutting down standard sizes, will vary depending upon your source of supply and how well you plan your project, but they cannot be neglected. Similarly, there are some items such as screws and paint that will be left over when the job is complete and, although usable in the future, they should be charged to the job. Lastly, you will make some mistakes in cutting or joining materials, resulting in scrap. As a rule of thumb based on my own experience, add 10 percent to the raw material cost to cover these factors.

PERFORMANCE TESTS

In the final analysis, it is not the cost per *square foot* of collector that determines economics, but the cost per *Btu* of energy collected and *used*. Simply stated, cost goes up and efficiency declines as you increase the collector temperature. This is illustrated in Fig. 4-1 and further explained in Table 4-1.

To get an efficient collector to handle a temperature difference of 200° F could require either a focusing collector or a flat plate type with a selective surface and a high vacuum between the absorber and cover pane (curve 4 in Fig. 4-1). These units can easily cost four times as much as a good unit having either one cover pane and a selective surface, or one having two covers and a flat-black paint coat on the absorber (curve 2). For the hot water and room preheaters described in this book, a collector operating on curve 1 or 2 will be all that is needed.

In summary: Don't pay for collecting higher temperatures than you need.

The analysis is easy to state, but it is difficult to quantify. Because the solar business is still in its infancy, collector manufacturers use different methods of developing and presenting efficiency curves for their products. The National Bureau of Standards has recognized this problem and has now developed a proposed test method for rating the thermal performance of solar collectors, namely test method NBSIR

Table 4-1. Conditions and Characteristics of Collectors Tested. Courtesy Arthur D. Little, Inc.

Normal Incident Radiation = 250 Btu/hr/sq ft
Ambient Air Temperature = 50°F
Spacing of Cover Panes = 0.5 in.

Curve	Collector Description	Number Of Panes	Absorber Properties	
			Absorptance	Emittance
1	flat black, unevacuated	1	0.93	0.93
2	flat black, unevacuated	2	0.93	0.93
3	selective, unevacuated	2	0.90	0.10
4	selective, high vacuum	1	0.90	0.10

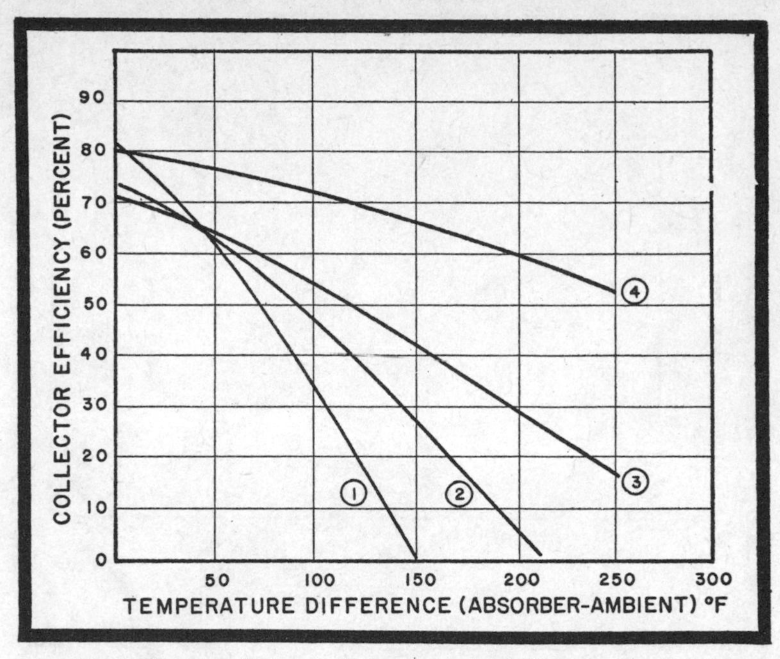

Fig. 4-1. Collector efficiency for various designs and operating temperatures.

74-635. Various testing laboratories throughout the country have been certified to conduct these tests on an impartial basis. Quality collector manufacturers will have their products independently tested by this method and publish the results. The curves developed from the NBS test are normalized for different *insolation* levels. (Insolation is the radiation from the sun.) By referring back to Fig. 4-1 and Table 4-1, you will see that the curves constructed were for a specified radiation level of 250 Btus per hour per square foot (Btu/hr-ft^2). Because this figure varies considerably from month to month by geographic location, dividing the temperature differential (ΔT) by the insolation level during testing gives a meaningful method of comparison without the need to draw curves for each and every insolation level you might encounter. A typical thermal efficiency curve for a flat plate collector tested by this method is shown in Fig. 4-2. Curves of this type (which are generally plotted as straight lines) provide an indication of the relative performance of different collector designs operating under *steady-state* test conditions. These conditions include:

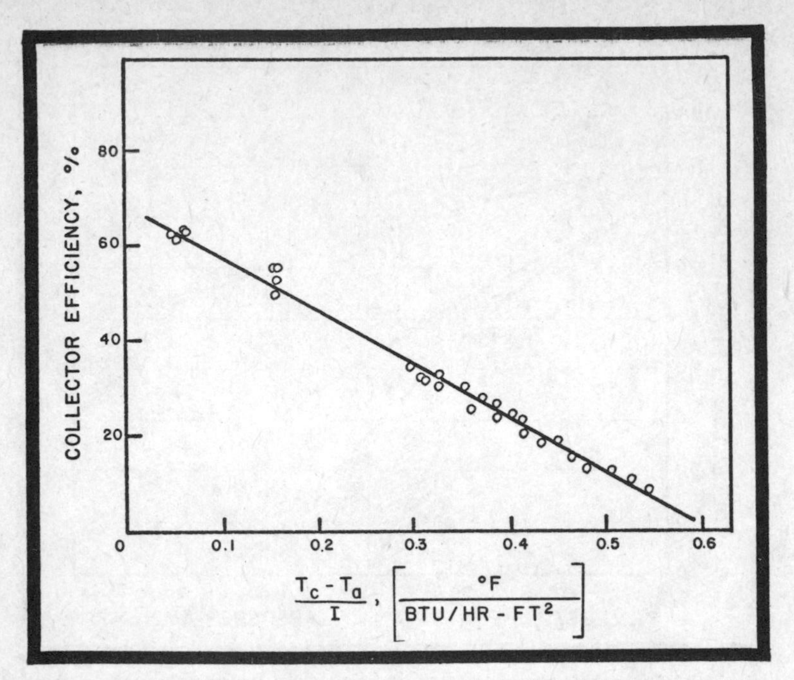

Fig. 4-2. Typical efficiency curve for a dual-glazed flat plate collector of the liquid type.

- fixed wind speed
- constant flow rate of liquid through the collector
- relatively constant insolation rate for the test period
- constant temperature of the liquid entering the collector

Under actual weather conditions (outdoor testing) both wind speed and insolation will vary, but the test method averages these values during the 15-minute testing period. These and other variables, such as the angle of the sun as it travels across the sky, result in the test points falling both above and below the curve plotted.

Now let's get down to some meaningful comparisons. Suppose you wish to compare the *cost effectiveness*—that is, the theoretical cost per Btu of energy collected—for three different collector designs. Assume that Collector A sells for $6.50/ft^2 ; Collector B, for $8/ft^2 ; and Collector C, for $10.50/ft^2 and that they have all been tested by the NBS method and plotted in Fig. 4-3. The application being considered is for residential hot water heating with an average desired *absorber temperature* (t_c) of 100° F. The location of the residence is in Grand Junction, Colorado, where the normal daily average temperature is 53° F, and we will assume this is the *ambient air temperature* (t_a) the collector is exposed to most of the year. Although more precise methods and tables for calculating *insolation* (I) levels are provided in the *Homeowner's Guide to Solar Heating & Cooling*, Table 4-2 provides an approximation by supplying the mean daily solar radiation in langleys for different parts of the United States. The average annual level for Grand Junction, Colorado is 456 langleys. Most of this falls in the 8-hour period from 8:00 a.m. to 4:00 p.m. Each langley provides 3.69 Btu, so the insolation value is approximately:

$$I = \frac{456}{8} \times 3.69 \qquad I = 210 \frac{\text{Btu}}{\text{hr}-\text{ft}^2}$$

This value is conservative because it neglects the increased insolation that a properly inclined collector intercepts.

Now the operating point on the efficiency curve (Fig. 4-3) can be calculated. The values on the horizontal axis are:

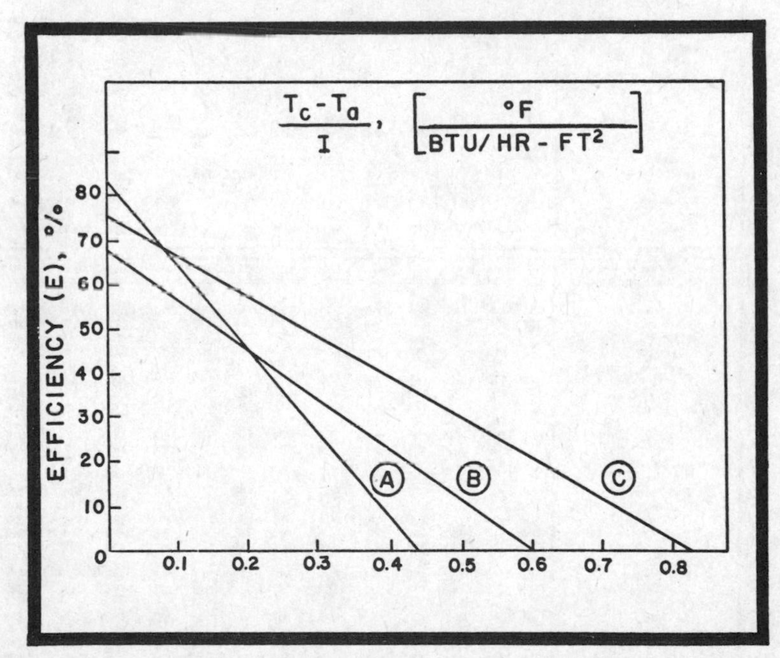

Fig. 4-3. Comparative efficiencies for three collector designs. T_c is absorber termperature, T_a is ambient air temperature and I is insolation.

Table 4-2. Mean Daily Solar Radiation (Langleys).

States and Stations	JAN	FEB	MAR	APR	MAY	JUNE	JULY	AUG	SEPT	OCT	NOV	DEC	ANNUAL
ALASKA													
Annette	63	115	236	364	437	438	438	341	258	122	59	41	243
Barrow	#	38	180	380	513	528	429	255	115	41	#	#	206
Bethel	38	108	282	444	457	454	376	252	202	115	44	22	233
Fairbanks	16	71	213	376	461	504	434	317	180	82	26	6	224
Matanuska	32	92	242	356	436	462	409	314	198	100	38	15	224
ARIZONA													
Page	300	382	526	618	695	707	680	596	516	402	310	243	498
Phoenix	301	409	526	638	724	739	658	613	566	449	344	281	520
Tucson	315	391	540	655	729	699	626	588	570	442	356	305	518
ARKANSAS													
Little Rock	188	260	353	446	523	559	556	518	439	343	244	187	385
CALIFORNIA													
Davis	174	257	390	528	625	694	682	612	493	347	222	148	431
Fresno	184	289	427	552	647	702	682	621	510	376	250	161	450
Inyokern (China Lake)	306	412	562	683	772	819	772	729	635	467	363	300	568
LaJolla	244	302	397	457	506	487	497	464	389	320	277	221	380
Los Angeles WBAS	248	331	470	515	572	596	641	581	503	373	289	241	463
Los Angeles WBO	243	327	436	483	555	584	651	581	500	362	281	234	436
Riverside	275	367	478	541	623	680	673	618	535	407	319	270	483
Santa Maria	263	346	482	552	635	694	680	613	524	419	313	252	481
Soda Springs	223	316	374	551	615	691	760	681	510	357	248	182	459
COLORADO													
Boulder	201	268	401	460	460	525	520	439	412	310	222	182	367
Grand Junction	227	324	434	546	615	708	676	595	514	373	260	212	456
Grand Lake (Granby)	212	313	423	512	552	632	600	505	476	361	234	184	417
DISTRICT OF COLUMBIA													
Washington (C.O.)	174	266	344	411	551	494	536	446	375	299	211	166	356
American University	158	231	322	398	467	510	496	440	364	278	192	141	333
Silver Hill	177	247	342	438	513	555	511	457	391	293	202	156	357
FLORIDA													
Apalachicola	298	367	441	535	603	578	529	511	456	413	332	262	444
Belle Isle	297	330	412	463	483	464	488	461	400	366	313	291	397
Gainsville	267	343	427	517	579	521	488	483	418	347	300	233	410
Miami Airport	349	415	489	540	553	532	532	505	440	384	353	316	451
Tallahassee	274	311	423	499	547	421	508	542	·	·	292	230	---
Tampa	327	391	474	539	596	574	534	494	452	400	356	300	453
GEORGIA													
Atlanta	218	290	380	488	533	562	532	508	416	344	268	211	396
Griffin	234	295	385	522	570	577	556	522	435	368	283	201	413
HAWAII													
Honolulu	363	422	516	559	617	615	615	612	573	507	426	371	516
Mauna Loa Obs	522	576	680	689	727	·	703	642	602	560	504	481	---
Pearl Harbor	359	400	487	529	573	566	589	567	539	466	386	343	484
IDAHO													
Boise	138	236	342	485	585	636	670	576	460	301	182	124	395
Twin Falls	163	240	355	462	552	592	602	540	432	286	176	131	378
ILLINOIS													
Chicago	96	147	227	331	424	458	473	403	313	207	120	76	273
Lemont	170	242	340	402	506	553	540	498	398	275	165	138	352
INDIANA													
Indianapolis	144	213	316	396	488	543	541	490	405	293	177	132	345
IOWA													
Ames	174	253	326	403	480	541	436	460	367	274	187	143	345
KANSAS													
Dodge City	255	316	418	528	568	650	642	592	493	380	285	234	447
Manhattan	192	264	345	433	527	551	531	526	410	292	227	156	371
KENTUCKY													
Lexington	172	263	357	480	581	628	617	563	494	357	245	174	411
LOUISIANNA													
Lake Charles	245	306	397	481	555	591	526	511	449	402	300	250	418
New Orleans	214	259	335	412	449	443	417	416	383	357	278	198	347
Shreveport	232	292	384	446	558	557	578	528	414	354	254	205	400
MAINE													
Caribou	133	231	364	400	476	470	508	448	336	212	111	107	316
Portland	152	235	352	409	514	539	561	488	383	278	157	137	350
MASSACHUSETTS													
Amherst	116	·	300	·	431	514	·	·	·	·	152	124	---
Blue Hill	153	228	319	389	469	510	502	449	354	266	162	135	328
Boston	129	194	290	350	445	483	486	411	334	235	136	115	301
Cambridge	153	235	323	400	420	476	482	464	367	253	164	124	322
East Wareham	140	218	305	385	452	508	495	436	365	258	163	140	322
Lynn	118	209	300	394	454	549	528	432	341	241	135	107	317
MICHIGAN													
East Lansing	121	210	309	359	483	547	540	466	373	255	136	108	311
Sault St Marie	130	225	356	416	523	557	573	472	322	216	105	96	333
MINNESOTA													
St Cloud	168	260	368	426	496	535	557	486	366	237	146	124	348
MISSOURI													
Columbia (C.O.)	173	251	340	434	530	574	574	522	453	322	225	158	380
Univ of Missouri	166	248	324	429	501	560	583	509	417	324	177	146	365
MONTANA													
Glasgow	154	258	385	466	568	605	645	531	410	267	154	116	388
Great Falls	140	232	366	434	528	583	639	532	407	264	154	112	366
Summit	122	162	268	414	462	493	560	510	354	216	102	76	312

Table 4-2. Cont'd.

States and Stations	JAN	FEB	MAR	APR	MAY	JUNE	JULY	AUG	SEPT	OCT	NOV	DEC	ANNUAL
NEBRASKA													
Lincoln	188	259	350	416	494	544	568	484	396	296	199	159	363
North Omaha	193	299	365	463	516	546	568	519	410	298	204	170	379
NEVADA													
Ely	236	339	468	563	625	712	647	618	518	394	289	218	469
Las Vegas	277	384	519	621	702	748	675	627	551	429	318	258	509
NEW JERSEY													
Seabrook	157	227	318	403	482	527	509	455	385	278	192	140	339
NEW HAMPSHIRE													
Mt. Washington	117	218	238	*	*	*	*	*	*	*	*	96	—
NEW MEXICO													
Albuquerque	303	386	511	618	686	726	683	626	554	438	334	276	512
NEW YORK													
Ithaca	116	194	272	334	440	501	515	453	346	321	120	96	302
N.Y. Central Park	130	199	290	369	432	470	459	389	331	242	147	115	298
Sayville	160	249	335	415	494	565	543	462	385	289	186	142	352
Schenectady	130	200	273	338	413	448	441	397	299	218	128	104	282
Upton	155	232	339	428	502	573	543	475	391	293	182	146	355
NORTH CAROLINA													
Greensboro	200	276	354	469	531	564	544	485	406	322	243	197	383
Hatteras	238	317	426	569	635	652	625	562	471	358	282	214	443
Raleigh	235	302	*	466	494	564	535	476	379	307	235	199	—
NORTH DAKOTA													
Bismarck	157	250	356	447	550	590	617	516	390	272	161	124	369
OHIO													
Cleveland	125	183	303	286	502	562	562	494	278	289	141	115	335
Columbus	128	200	297	391	471	562	542	477	422	286	176	129	340
Put-in-Bay	126	204	302	386	468	544	561	487	382	275	144	109	332

States and Stations	JAN	FEB	MAR	APR	MAY	JUNE	JULY	AUG	SEPT	OCT	NOV	DEC	ANNUAL
OKLAHOMA													
Okla. City	251	319	409	494	536	615	610	593	487	377	291	240	436
Stillwater	205	289	390	454	504	600	596	545	455	354	269	209	405
OREGON													
Astoria	90	162	270	375	492	469	539	461	354	209	111	79	301
Corvallis	89	*	287	406	517	570	676	558	397	235	144	80	—
Medford	116	215	336	482	592	652	698	605	447	279	149	93	389
PENNSYLVANIA													
Pittsburgh	94	169	216	317	429	491	497	409	339	207	118	77	280
State College	133	201	295	380	456	518	511	444	358	256	149	118	318
RHODE ISLAND													
Newport	155	232	334	405	477	527	513	455	377	271	176	139	338
SOUTH CAROLINA													
Charleston	252	314	388	512	551	564	520	501	404	338	286	225	404
SOUTH DAKOTA													
Rapid City	183	277	400	482	532	585	590	541	435	315	204	158	392
TENNESSEE													
Nashville	149	228	322	432	503	551	530	473	403	308	208	150	355
Oak Ridge	161	239	331	450	518	551	526	478	416	318	213	163	364
TEXAS													
Brownsville	297	341	402	456	564	610	627	568	475	411	296	263	442
El Paso	333	430	547	654	714	729	666	640	576	460	372	313	536
Ft. Worth	250	320	427	488	562	651	613	593	503	403	306	245	445
Midland	283	358	476	550	611	617	608	574	522	396	325	275	466
San Antonio	279	347	417	445	541	612	639	585	493	398	295	256	442

Langley is the unit used to denote one gram calorie per square centimeter.

States and Stations	JAN	FEB	MAR	APR	MAY	JUNE	JULY	AUG	SEPT	OCT	NOV	DEC	ANNUAL
UTAH													
Flaming Gorge	238	298	443	522	565	650	599	538	425	352	262	215	426
Salt Lake City	163	256	354	479	570	621	620	551	446	316	204	146	394
VIRGINIA													
Mt. Weather	172	274	338	414	508	525	510	430	375	281	202	168	350
WASHINGTON													
North Head	*	167	257	432	509	487	486	436	321	205	122	77	—
Friday Harbor	87	157	274	418	514	578	586	507	351	194	102	75	320
Prosser	117	222	351	521	616	680	707	604	458	274	136	100	399
Pullman	121	205	304	462	558	653	699	562	410	245	146	96	372
Univ. of Washington	67	126	245	364	445	461	496	435	299	170	93	59	272
Seattle-Tacoma	75	139	265	403	503	511	566	452	324	188	104	64	300
Spokane	119	204	321	474	563	596	665	556	404	225	131	75	361
WISCONSIN													
Madison	148	220	313	394	466	514	531	452	348	241	145	115	324
WYOMING													
Lander	226	324	452	548	587	678	651	586	472	354	239	196	443
Laramie	216	295	424	508	554	643	606	536	438	324	229	186	408
ISLAND STATIONS													
Canton Island	588	626	634	604	561	549	550	597	640	651	600	572	597
San Juan, P.R.	404	481	580	622	519	536	639	549	531	460	411	411	512
Swan Island	442	496	615	646	625	544	588	591	535	457	394	382	526
Wake Island	438	518	577	627	642	656	629	623	587	525	482	421	560

* insufficient data collected
— incomplete monthly data
Barrow is in darkness during the winter months

$$\frac{t_c - t_a}{I} \left[\frac{°F}{[Btu/hr - ft^2]}\right]$$

where:

t_c = absorber temperature
t_a = ambient air temperature
I = insolation

$$\frac{100 - 53}{210} = 0.22$$

By drawing a vertical line through the three performance curves at this point, the efficiencies of the three collectors being considered may be read from the chart. Looking at Fig. 4-4, the following efficiencies may be read:

Collector	Efficiency
A	0.42
B	0.44
C	0.55

The useful heat output from the collector (Q_u) may be estimated by multiplying the efficiency (E) by the insolation (I).

$$Q_u = E\,I \frac{Btu}{[hr - ft^2]}$$

When the collector cost is divided by Q_u, the result is a value relating cost to its energy-producing potential. Let's call this value the *effective cost*.

Collector	Effective Cost	$\frac{\text{Collector Cost/ft}^2}{Q_u}$
A	88.2	.07
B	92.4	.09
C	115.5	.09

This *effective cost* cannot be equated to any fuel costs. It is merely a *guide* to the best collector value under the operating conditions chosen. Strictly on test data, which measures the steady-state performance, Collector A is the best buy.

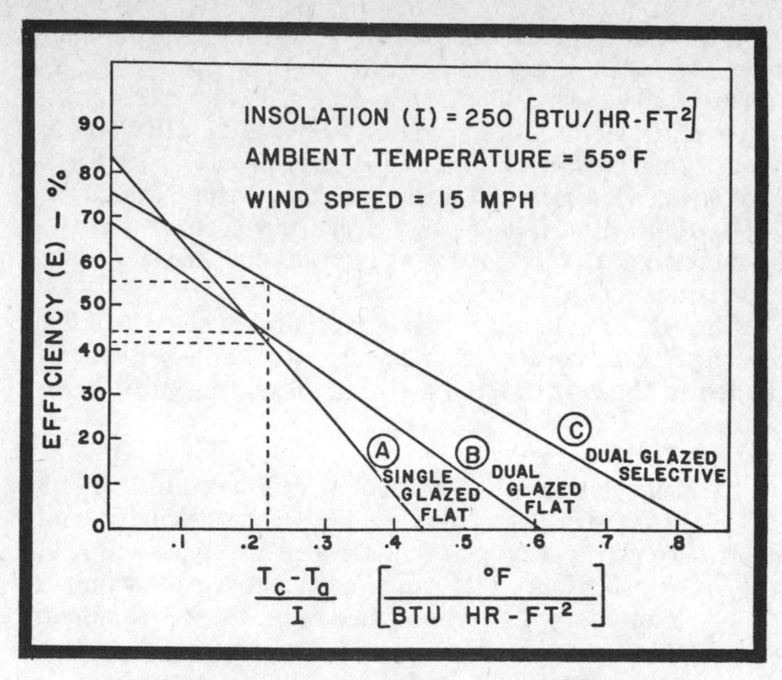

Fig. 4-4. Comparative efficiencies at a selected operating point.

There are other factors to consider, however, because the response time of two units may vary considerably. For example, if the absorber of the collector is constructed of thin-gauge copper, it will reach operating temperature much faster than a unit made of steel, which has less conductivity. The copper unit has a definite advantage in areas with intermittent sunshine. If you have steady sunshine throughout the day, it doesn't make much difference. The steel warms more slowly, but it will hold its heat longer. Aluminum is a compromise between copper and steel.

GUARANTEES

If the solar collectors fail prematurely (that is, before the duration of their intended design life), you will be the loser unless a warranty is available and adjustments are made. There are, of course, no warranties either expressed or implied to cover the performance, materials, or workmanship on collectors you build yourself. But when you invest in a factory-made unit, you are paying for, and should receive, some assurance that the product will work.

Both government and industry leaders are working diligently to define solar standards. I recently was asked to review a document which, when finalized, will become the bench mark of the solar industry. This is NBSIR 76-1059, entitled *Intermediate Minimum Property Standards for Solar Heating and Domestic Hot Water Systems*. It was prepared for the Department of Housing and Urban Development (HUD), Washington, D.C. 20410, and a copy may be obtained by writing to the department.

The HUD minimum property standards require both the labeling and testing of solar collectors. Applicable sections, as printed in the April 1976 issue of the document, are shown below.

S-515-2.1.1 Labeling—

Collectors shall be labeled in accordance with 515-1.2 of the MPS. In addition, collector labels shall list total weight, cover plate materials, the types of heat transfer fluids that can or cannot be used, maximum allowable operating and no-flow pressures, maximum flow rates, and collector efficiency as measured according to S-615-2.2.1.

S-515-1.2 Temperature and Pressure Resistance—

Components shall be capable of performing their functions for their intended design life when exposed to the temperatures and pressures that can be developed in the system under both flow and no-flow conditions.

S-615-2.2.1(a) Flat Plate Collectors—

This type collector can be characterized as a subsystem or component for an active solar system requiring the use of powered mechanical equipment to move the heat transfer fluid (liquid or gas) through the collector. The collector thermal performance shall be based upon the slope-intercept method of expressing efficiency for the range of operating conditions including solar power density, heat transfer fluid temperature, ambient temperature, wind, solar radiation incident angle, and flow rates, to be used in the design.

Commentary

The collector performance characteristics can be measured using the National Bureau of Standards proposed test method (NBSIR 74-635) for rating solar collectors or any other method demonstrated to have an overall limit-of-error of less than ±5 percent. This method provides sufficient efficiency versus operating condition data to construct a curve normalized for insolation and temperature difference between ambient and heat transfer fluid temperature. Curves for typical flat-black and selective coated absorber panels with one and two covers are shown in Appendix A for air and water collectors. Collectors with other geometric, optical, or thermal characteristics may require additional tests to fully describe their thermal performance for all environmental and operating conditions. The operating electrical power is recorded and reported during all tests.

Look to the seller of a commercial collector for a guarantee or warranty that the unit will perform under the stated conditions for a specified period of time.

If you elect to build your own solar collectors, be sure to consider temperature and pressure resistance, durability, and theoretical performance. Because it is impractical to have do-it-yourself units tested commercially, your skill and know-how will determine the end result.

APPEARANCE

The collector designs suggested in this book have been carefully chosen to provide an overall appearance comparable to commercially manufactured units. In many cases, it is not the collectors themselves that people object to, but the way they are mounted, spaced, or connected using a large quantity of exposed hoses and pipes. These problems can be minimized by careful planning and installation.

For example, good design calls for the collectors to be held above the roof surface with enough air space underneath them to prevent trapped water from causing roof rot, fungus, mold, mildew, and leakage. If the inlet and outlet fittings are properly located, feed and return lines to the panels may be concealed *underneath* the assembly with access to the connections still maintained for servicing. One trick is to conceal the vertical pipes running to and from the collectors in a downspout along the corner of the house.

Another consideration may be the reflection of sunlight or glare from glass covers used on the collectors. If this is of concern to you or to your neighbors, you can buy varieties of low-reflectance glass (but they are expensive) or fiberglass-reinforced plastic that may be used as collector covers if you build your own. Some commercial units use an acrylic plastic dome as a cover, to help the glare problem.

COLLECTOR PANEL SIZE

There are a number of factors you should consider in comparing the panel sizes of commercially available collectors with those of the one you can construct yourself. First, since each panel is self-contained in a frame, whenever you can reduce the overall perimeter of the installation you should achieve a corresponding savings in material costs. Secondly, heat losses along the edges of the collector are greater in small units where the *perimeter* is large in comparison with the *area* of the collector panel. And lastly, it takes more connections, hoses, fittings, and mounting brackets to install a number of smaller units than one larger unit. Once you have selected specific collectors and have determined the total area of panels required for your project, cost comparisons can be made.

The limiting factors in the size of commercial collectors generally relate to the choice and availability of cover materials and absorbers. For example, tempered glass is a popular option for covers because it resists vandalism, hailstones, etc. But tempered glass cannot be cut after it is heat treated or tempered, so some glass producers select the most popular size, thereby guaranteeing high volume; they then price that size competitively. Other sizes may be available through specialty suppliers, but only at high prices. The size preferred by the glass industry is used in the manufacture of standard size patio doors, which are approximately 34″ × 76″. Cover materials such as fiberglass-reinforced plastics (0.040 inch thick) or thin plastic films (0.004-inch thick polyvinyl fluoride) are available in larger sizes. Sandwich-type absorber panels made by the Olin Brass Roll-Bond process are popular among collector manufacturers, but again there are size limitations. Olin's literature indicates the size capabilities, shown in Table 4-3.

One last comment on size: Be sure to compare units based on their *effective absorber area* and not their overall size. The

Table 4-3. Size Limitations for Olin Roll-Bond Solar Absorbers.

	Thickness*	Width	Length
Copper	.040″ min.	34″ max.	96″ max.
Aluminum	.060″ max.	36″ max.	110″ max.

*Tube wall is approximately 1/2 sheet thickness.

frame around a collector, for example, does not make any contribution to the solar energy you are trying to collect. Similarly, in mounting the absorber, part of its surface is either concealed or otherwise shaded from the sun. A quality manufacturer will provide not only overall dimensions of his unit, but also the dimensions of the effective absorber area in square feet; or, he will otherwise state that the published performance of the unit is based on the gross area of the collector, if this is the case.

BUILDING CODES

Many people confuse *codes* with *standards*. The National Bureau of Standards (NBS) explains the difference this way: standards provide the basis for mass production and interchangeability of building components. Codes utilize standards as the basis for identifying the minimum acceptable level of both the quality and performance of the component. Once adopted, a code then becomes a legal method of regulation by which a city, county, or state may control the practice, design, and installation of the building component. This regulation is not arbitrary but is predicated on the need for the protection of public health and safety. Environmental conditions vary throughout the United States, so modifications in a model code become necessary to compensate for local conditions such as earthquakes, hurricanes, freezing, or differences in soils and water supplies.

Codes generally apply to existing standards, but in the case of solar energy, the field is so new that standards have not been finalized. Earlier in this chapter, I referenced the best overall document the industry has today—*Intermediate Minimum Property Standards for Solar Heating and Domestic Hot Water Systems.* There are some sections of this document that should be helpful to you in making the decision on whether to build or buy a collector. I previously mentioned sections S-515-2.1.1, S-515-1.2, and S-615-2.2.1(a) relating to labeling, temperature and pressure resistance, and testing of flat plate collectors. In addition, here are some other key sections for your consideration:

S-501-1 Materials—

Materials installed shall be of such kind and quality as to assure that the solar energy system will provide (a) adequate structural strength, (b) adequate resistance to weather, moisture, corrosion, and fire, (c) acceptable durability and economy of maintenance and market acceptance.

S-501-3.1 Protection of potable water and circulated air—

No material, form of construction, fixture, appurtenance, or item of equipment shall be employed that will support the growth of micro-organisms or introduce toxic substances, impurities, bacteria, or chemicals into potable water and air circulation systems in quantities sufficient to cause disease or harmful physiological effects.

Commentary

This situation is of concern not only as it pertains to ducts, piping, filters, and joints but also to storage areas, such as rock beds. In addition, the growth of fungus, mold, and mildew is possible when collectors are applied to a roof surface over the water-tight membrane. If the collectors are in contact with the membrane or held away from the membrane to allow for drainage, the shaded membrane area can support the growth of mildew and other fungus in some warm, moist climates. Special design considerations should be included to avoid this problem in climates where it can occur.

S-515-1.3 Materials Compatibility—

All materials which are joined to or in contact with other materials shall have sufficient chemical compatibility with those materials to prevent deterioration that will significantly impair their function during their intended design life. Provisions shall be made to allow for differences in the expansion of joined materials.

S-515-2.1.3 Flashing—

a. Flashing for collector panel supports that penetrate the primary roof membrane shall be designed to prevent the penetration of water or melting snow for the life of the roof system.
b. Flashing systems shall be designed to permit minor repairs without disturbing the roof membrane, collector supports, or collector panels.
c. In general, flashing for roof penetration shall comply with applicable Sections of 507-5 and 507-8 of the MPS.

Commentary
Suggested practices for flashing used on no-slope or low-slope roofs and roof penetrations are provided in the National Roofing Contractors Association's *A Manual of Roofing Practice,* 1970.

S-151-2.2.4 Structural Requirements—
All glazing materials shall be of adequate strength and durability to withstand the loads and forces required by Section S-601 of this document.

S-600-6.1 General—
Materials or construction used in the installation of solar systems shall be in accordance with the fire protection provisions of Section 405-7 of MPS.

Commentary (W.M. Foster): The MPS 4900, or *Minimum Property Standards for One and Two Family Dwellings*, is available from HUD regional offices. Chapter 6 (Section 601) of the *Intermediate Minimum Property Standards for Solar Heating and Domestic Hot Water Systems* provides complete details on the structural requirements solar collectors must meet when either roof- or ground-mounted. Such factors as wind loads, snow loads, and hail loads vary widely for each geographic region. I would suggest that you obtain a copy of this document from HUD to check your local conditions. In addition, there are several pages relating to the compatibility of the collector absorber plate and the heat transfer fluid used in the solar system. This commentary is too extensive to include here, but the general conclusion is that copper absorbers are suitable for circulating potable water *if* the water is low in carbonates and chlorides, or de-ionized, and the velocity of the water through the absorber tubes is less than 4 feet per second.

There are four basic code-writing bodies in the U.S. today:

ICBO—International Conference of Building Officials (Uniform Building Code).

SBCC—Southern Building Code Congress International, Inc. (Standard Building Code).

BOCA—Building Officials and Code Administrators International, Inc. (Basic Building Code).

AINA—American Insurance Association (National Building Code).

I used the word *basic* because there are over 1000 different building codes in existence. This fragmentation has plagued our country in implementing worthwhile changes in construction methods. In the case of solar energy, it is hoped that coordination on a national level will result in consensus standards being accepted uniformly as a supplement to building codes. Generally, you will find that ICBO is used on the West Coast and in the central Midwest, SBCC in the South, and BOCA in the upper Midwest, New England, and the Mid-Atlantic States. When new building products are developed, manufacturers usually seek approval from a code body active in the area where they intend to market the product. Once approvals are secured, circulars are sent to local building departments who generally adopt the code body's recommendation. For products you assemble, it will be up to *you* to convince your local building department that your plans, assembly, and installation meet its specifications.

Selecting A System to Suit Your Climate

From a load standpoint, there are two separate analyses that must be made in selecting a system to suit your climate: one for a solar-assisted water heater with an almost constant load throughout the year, and one for a room heater with 80–90 percent of the load occuring during the heating season (November to April).

Because *solar fuel* or sunshine is relied upon to satisfy a portion of the load, this commodity must be available and be reasonably predictable during the peak load period. Also, local climate conditions such as freezing weather, hail, wind, and frequent fog must all be considered when selecting a system.

HOT WATER HEATERS

Let's start with the solar hot water heater having a relatively constant load. There are three system types to consider:

- thermosiphon
- pumped system circulating potable water
- closed-loop heat exchanger system

In the first two cases, freezing weather can present problems unless the systems are drained. A closed loop that circulates antifreeze solution does not have this problem, but its heat exchanger adds cost and lowers efficiency. The ideal situation is to have maximum daily solar radiation over the entire year. Those zones bordered by the 500-langley lines in Fig. 5-1 would therefore be considered optimum areas for solar hot water heaters. Also, an area where freezing weather does not occur

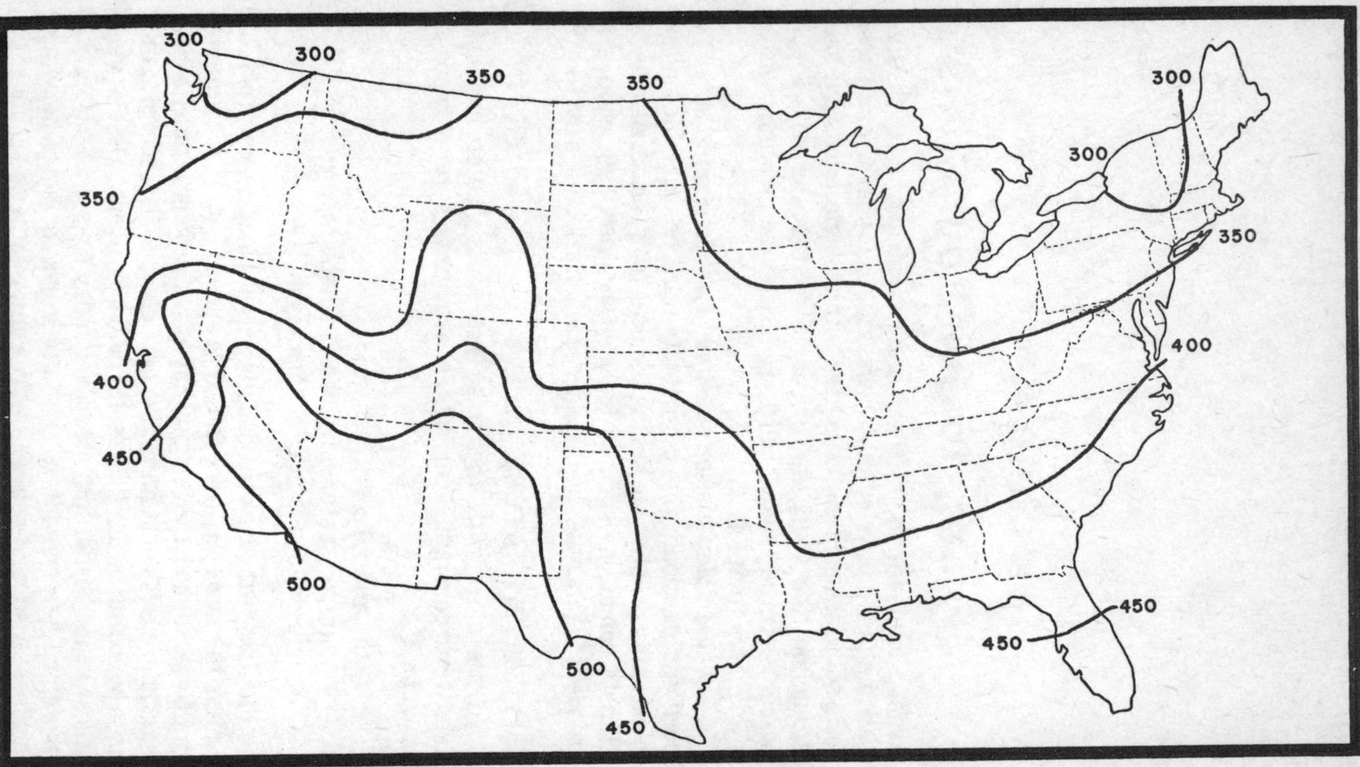

Fig. 5-1. Mean daily solar radiation (langleys)—annual.

at all, such as in Hawaii, or occurs infrequently, (say, two times per year), is advantageous from a cost standpoint, because an inexpensive thermosiphon system may be used in such areas. When a freeze is predicted, the collectors can be manually drained. In Fig. 5-2, you will see areas with infrequent freezes designated zone A. For areas having more frequent freezes, it is practical to consider a pumped system circulating tap water which automatically drains the exposed plumbing when freezing weather occurs. Every time the system *dumps* or drains, you lose roughly three gallons of water; but because this design operates more efficiently and at lower initial cost than a unit having a heat exchanger, it is recommended for homeowners living in zones B and C, where freezes are not overly frequent.

Most of the United States lies in zone D where freeze-ups occur quite frequently, however. Automatic valves for drainage are less reliable in cold climates, and the loss of water each time the system drains will start to add up. For these reasons, I recommend that zone D residences use a design circulating a fluid with a low freezing point through the collectors in a closed loop, and use a heat exchanger to warm the potable water in the storage tank. Such a system is covered in Chapter 8.

SPACE OR ROOM HEATING

With the load on this space or room-heating system concentrated in the November through April period, the *annual* solar radiation plot in Fig. 5-1 cannot be used. Instead, solar radiation for the six-month *heating season* had to be computed and plotted on a similar map, again using 50-langley intervals. This provides a measure of the sunshine or solar fuel that can be expected for each zone we define. Because the amount of cold weather or *load* varies throughout the country, this load is the next variable that must be calculated. A method commonly used is called the *heating degree day*, which is the number of degrees below 65° F the daily average temperature reaches during a given season. Research has shown that home heaters will be used when the outdoor temperature drops below 65° F. The farther below 65° F the temperature falls, the more heat will be required to make a house comfortable. In Boston the average daily temperature is 44.9° F during November. This figure, the average daily

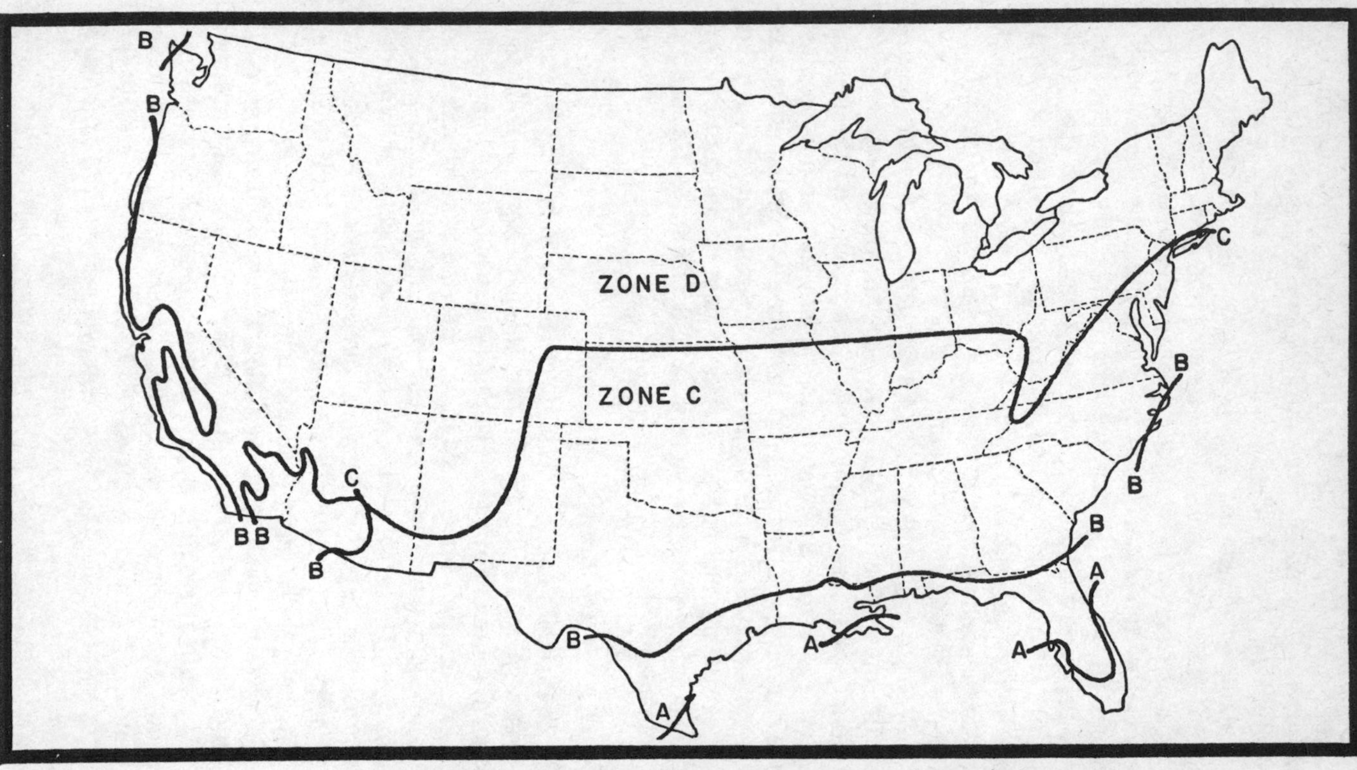

Fig. 5-2. Climatic zones suggested for various solar water heater designs.

temperature, is subtracted from 65°F to give the daily degree day reading. In this case, 20.1 multiplied by 30 (the number of days in November) provides the total heating degree days for Boston for the month: 603. Adding the monthly readings for the November—April period totals 5005. A similar calculation for Los Angeles would yield 1652, telling us that it requires three times as much fuel to heat a Boston residence as it would if the same structure were located in Los Angeles.

Because the degree day concept has been used in the heating industry for thirty years, substantial data are available to plot the lines of constant values across the United States. This work was done by the TRW Systems Group in its report to the National Science Foundation. The method used by TRW was to combine radiation and heating degree day data on a single map defining nine climate zones as a function of supply (sunshine) and demand (degree days). The result is shown in Fig. 5-3; but in using it, you should be aware that generalizations have been made in its compilation. For example, no corrections were made for topography, but you would expect an increase in both supply and demand as elevation increases. The method does provide a general guide, though, and Fig. 5-3 will be used as a reference for space or room heating climatic zones. Table 5-1 describes the insolation and degree day ranges for each zone, in addition to providing a description of heating requirements.

Only one space or room heating design is provided in this book, and it will not be cost effective in all zones. The determination of whether to use it must be made based upon your specific requirements, including local fuel costs, and I

Table 5-1. Parameters Used to Select Climatic Zones in Fig. 5-3.

Zone	Solar Energy Availability	Mean Daily Insolation (Langley/Day)	Heating Requirement	Heating Degree Days
1	Highest	350 – 450	Low to Moderate	0 – 2500
2	Highest	350 – 450	Moderate to High	2500 – 5000
3	Highest	350 – 450	High to Very High	5000 – 9000
4	Moderate	250 – 350	Low to Moderate	0 – 2500
5	Moderate	250 – 350	Moderate to High	2500 – 5000
6	Moderate	250 – 350	High to Very High	5000 – 9000
7	Lowest	175 – 250	Low to Moderate	0 – 2500
8	Lowest	175 – 250	Moderate to High	2500 – 5000
9	Lowest	175 – 250	High to Very High	5000 – 9000

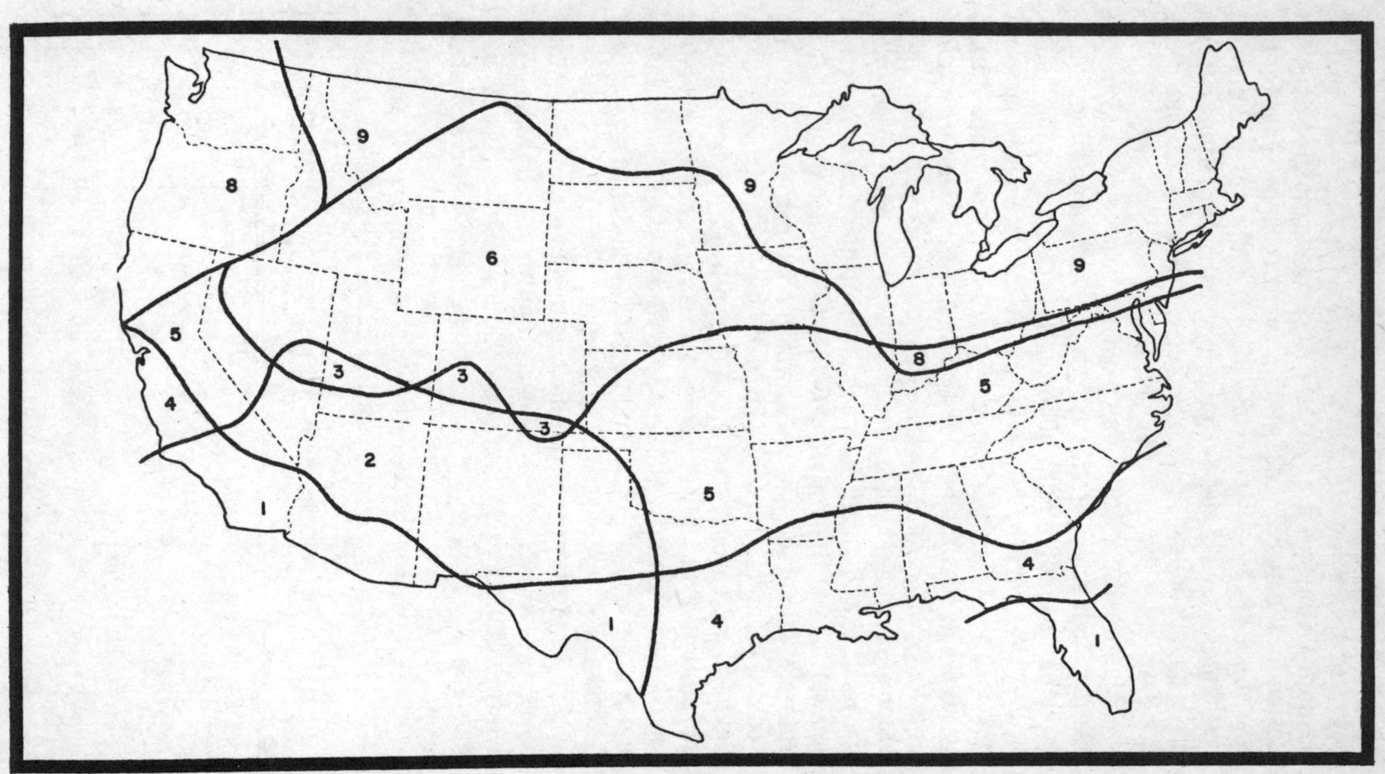

Fig. 5-3. Climatic zones for solar space heating. Source: TRW Systems Group.

would refer you to my *Homeowner's Guide to Solar Heating & Cooling* for methods of evaluation. From a strictly technical standpoint, the heating system of Chapter 9 should be considered for use in zones 2, 3, 5, and 6—or about half of the United States.

LOCAL CLIMATE CONDITIONS

The *quality* of insolation or solar radiation which you hope to capture with your solar system varies substantially throughout the United States. If you could collect the sunlight *before* it entered the earth's atmosphere, approximately 427 Btus per square foot per hour would be available. But as this radiation passes through the atmosphere it loses intensity due to the orientation of the earth, and the absorption and scattering of radiation by clouds, vapor, dust, smoke, smog, and other impurities. The influence of these factors is greater in winter months and northern latitudes; the sun travels lower in the sky and must pass through more of the atmosphere to reach these areas at these times. Also, you would expect increased pollution near large cities or industrial complexes. Readings can vary as much as 20 percent within a few miles, and unfortunately the weather stations with good, long-term data are widely dispersed. I mention these variables so you will not put blind faith in the insolation data provided by the U.S. Department of Commerce and tabulated in Table 5-1. These data may serve as a guide until more complete localized data are developed, but failure to recognize local factors could yield disappointing results.

The Path of the Sun

The earth travels *around* the sun, completing its orbit once each year. During this orbit, the earth *tilts* 23.5°, producing two equinoxes and two solstices each year. An *equinox* is defined as the time when the sun crosses the equator, making night and day of equal length in all parts of the world.

The spring equinox occurs on March 21 and the fall equinox, on September 21. A *solstice* occurs when the sun is at its maximum distance from the equator. The summer solstice is June 21 and the winter solstice, December 21. In our hemisphere, the sun is at its highest azimuth, or angle above the horizon, on June 21, and this is the *longest* day of the year. Conversely, on December 21 we have the *shortest* day.

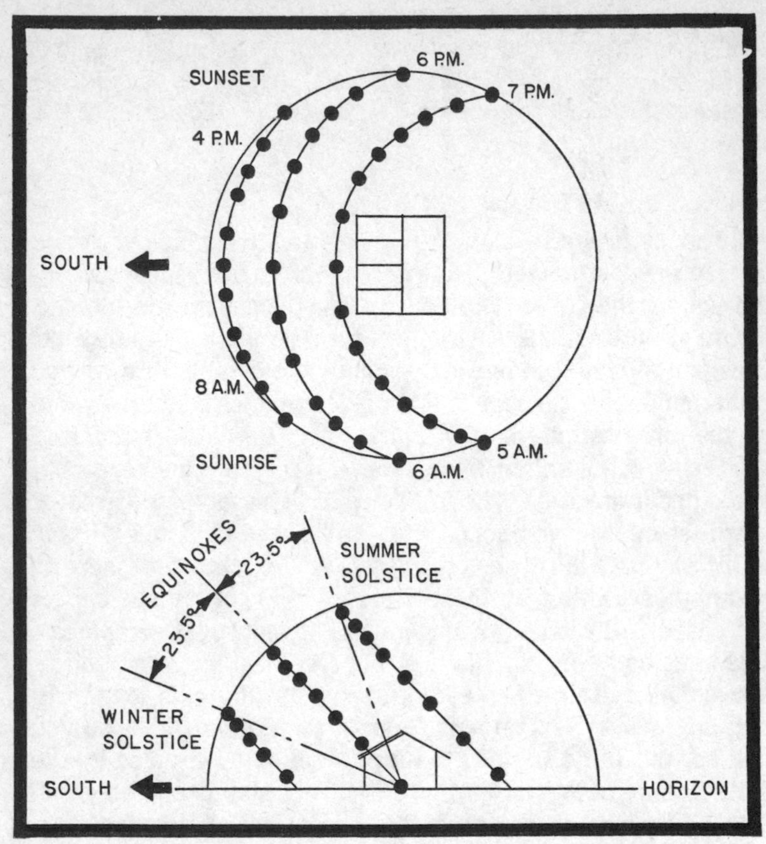

Fig. 5-4. Sun positions for the solstice and equinox periods.

Figures 5-4 and 5-5 may help you to understand these principles and their significance. In Fig. 5-4, the hourly position of the sun from sunrise to sunset is represented by a black dot. The bottom diagram shows that during the winter solstice the sun travels in a *low* arc close to the horizon. The further north you live, the lower this arc becomes. During the summer solstice, the sun travels much *higher* in the sky (as much as 47 degrees;), yielding a much longer day. The top portion of Fig. 5-4 shows this sun path from a different perspective, looking down on top of the solar house. For this example, the sun rises above the horizon at 8:00 a.m. and sets at 4:00 p.m. in the winter. For the summer solstice, sunrise is at 5:00 a.m. and sunset, 7:00 p.m.

The position of the sun for any day and hour of the year may be obtained from almanacs in terms of the altitude (A) and azimuth (B) angles. At 40° latitude on a winter solstice day the sun's *altitude* (A), or angular height above the horizon, at 9:00 a;m., is 14°. The corresponding azimuth (B), or angular distance from south, is 42°. Figure 5-5 shows how these angles are measured. Knowing the path of the sun will help you to understand collector orientation and shading factors which will be covered next.

Orientation of the Solar Collector

From the previous section, you should now be able to understand why it is necessary to incline or tilt the collectors. To intercept a maximum amount of sunshine, collectors should theoretically be perpendicular to the sun's rays, but because the sun changes its position daily, the most practical solution is to optimize the inclination for the *season* when the heating load is maximum, and mount the collectors in this *stationary* position.

In the case of a room heater, I have defined the *heating season* as October through April. I would suggest the collectors be inclined at an angle equal to your latitude plus 10° above horizontal. If you refer to Fig. 5-6, you will appreciate the importance of tilting the collectors to maximize winter sunshine; the further north you are, the more critical this becomes. For example, if your collectors were not inclined (that is, if they are horizontal), all the radiation you could expect to receive would be that shown in Table 5-1 which is for horizontal surfaces. But if you tilt the collectors to your

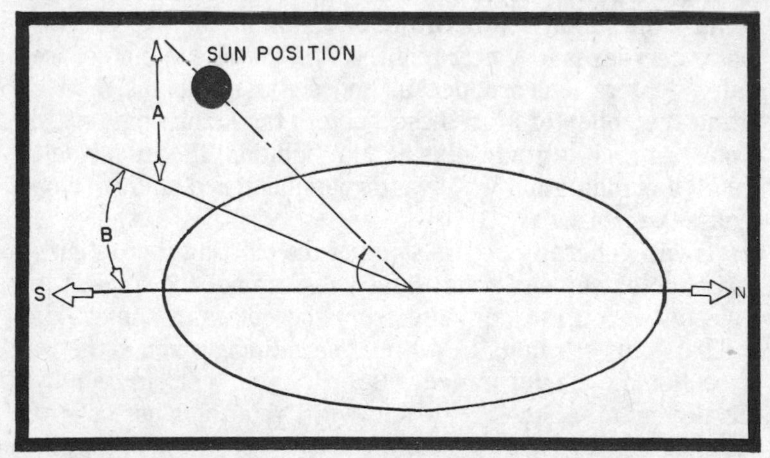

Fig. 5-5. Locating the sun position by altitude and azimuth angles.

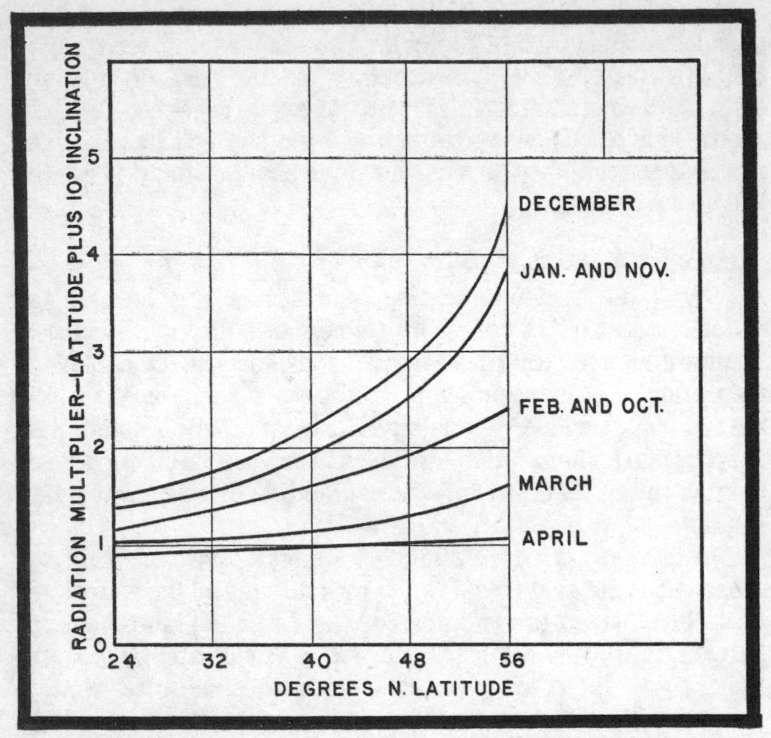

Fig. 5-6. Effect of collector tilt as a function of latitude and month.

latitude plus 10°, this value may be *multiplied* by the factor shown in Fig. 5-6.

Solar water heaters have a year-round heating load, but because the supply temperature of the water coming from city mains decreases in winter months, and because in many cases water heaters overproduce in hot, summer months, I favor tilting the collectors for these heaters the same angle as for room heating—latitude plus 10°. In addition, if you decide to install a combination water and room heater, a uniform slope makes the job easier.

It will generally be the slope of the existing roof intended for collector-mounting that finally determines the slope of the collectors. You may deviate from the optimum angle by as much as plus or minus 15° without sacrificing much in the way of collector performance. But if you get into *major* misalignments, like 45°, the sun will strike the glass cover on the collector at such a low angle that much of the energy will be reflected, causing a major loss in performance.

For ideal solar collection, the sloping collectors should face due south for all heating applications; but very few roof structures have this ideal orientation. So, as in the case above, we need to define the penalty for deviating from the optimum. A collector orientation of 20° on either side of true south is acceptable and may even be preferable depending upon your local conditions. For example, if you get afternoon shade, bias the collectors toward the east. Or, if morning fog is characteristic of your region, facing the collectors 10° to 20° west of south would be preferable. Performance falls off rapidly as you approach a 45° deviation. Variations in orientation are more critical in winter months and as latitude increases.

Shading Factors

A solar collector located in the shade is virtually useless, and yet many people find that the nice sunny roof they looked at in the summer is actually in the shade in winter months when the heating load is highest. This is a pitfall you can avoid with the knowledge of how the sun travels throughout the seasons, coupled with some basic measurements and calculations.

Shadows may be cast from portions of your own house—a stepped roof, chimney, or gables—as shown in Fig. 5-7, or from neighboring buildings, hills, and trees (Fig. 5-8). A foolproof

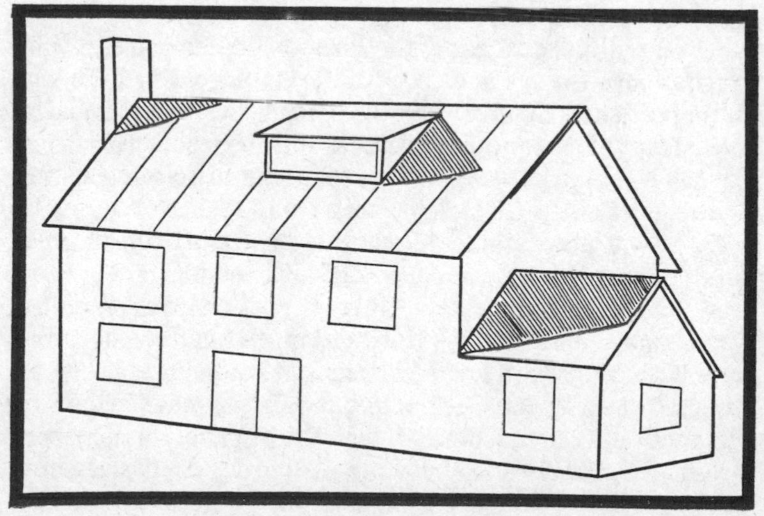

Fig. 5-7. Sources of shading from building elements.

Fig. 5-8. Shading effect of neighboring structures and landscape.

way to measure shading factors is to actually mark off the area where collectors are to be placed and observe the sunlight throughout the season solar collection is desired. Collection hours vary from 8:00 a.m. to 4:00 p.m. in the summer to 9:00 a.m. to 3:00 p.m. in the winter; this is generally the time that shade should be avoided. It may not be practical for you to use this year-long observation method, however. If that is the case, good approximations may be made by calculations.

You will need to know the altitude, or the sun's angular height above the horizon (see Fig. 5-5A), throughout the year during collection hours. Table 5-2 lists this value for 32° and 40° latitudes. Values for other latitudes may be found in almanacs or may be obtained by writing to the Defense Mapping Agency Hydrology Center, Clearfield, Utah 84016.

Now for an example of how these altitude angles may be used: suppose that your collectors are mounted on the house roof facing south, but the neighbor's house is 35 feet south of yours, and is 20 feet high. If the bottom edge of the collectors is 10 feet above ground level, will the sun shade the collectors in December at 40° latitude? Refer to Fig. 5-9 which shows the distance between objects as X feet, the height of the neighbor's house as H, the distance from the bottom edge of the collectors to the ground as h, and the angle formed from lines extending perpendicular to the ground, and another from the bottom

Table 5-2. Sun Positions for 32° and 40° North Latitudes.

DATA	SOLAR TIME		SOLAR ALTITUDE IN DEGREES	
	A.M.	P.M.	32° LATITUDE	40° LATITUDE
JAN 21	8	4	12.5	8.1
	9	3	22.5	16.8
	10	2	30.6	23.8
	11	1	36.1	28.4
	12		38.0	30.0
FEB 21	8	4	19.0	15.4
	9	3	29.9	25.0
	10	2	39.1	32.8
	11	1	45.6	38.1
	12		48.0	40.0
MAR 21	8	4	25.1	22.5
	9	3	36.8	32.8
	10	2	47.3	41.6
	11	1	55.0	47.7
	12		58.0	50.0
APR 21	8	4	31.5	30.3
	9	3	43.9	41.3
	10	2	55.7	51.2
	11	1	65.4	58.7
	12		69.6	61.6
MAY 21	8	4	35.4	35.4
	9	3	48.1	46.8
	10	2	60.6	57.5
	11	1	72.0	66.2
	12		78.0	70.0
JUN 21	8	4	36.9	37.4
	9	3	49.6	48.8
	10	2	62.2	59.8
	11	1	74.2	69.2
	12		81.5	73.5
JUL 21	8	4	35.7	35.8
	9	3	48.4	47.2
	10	2	60.9	57.9
	11	1	72.4	66.7
	12		78.6	70.6
AUG 21	8	4	31.8	30.7
	9	3	44.3	41.8
	10	2	56.1	51.7
	11	1	66.0	59.3
	12		70.3	62.3
SEPT 21	8	4	25.1	22.5
	9	3	36.8	32.8
	10	2	47.3	41.6
	11	1	55.0	47.7
	12		58.0	50.0
OCT 21	8	4	18.7	15.0
	9	3	29.5	24.5
	10	2	38.7	32.4
	11	1	45.1	37.6
	12		47.5	39.5
NOV 21	8	4	12.7	8.2
	9	3	22.6	17.0
	10	2	30.8	24.0
	11	1	36.2	28.6
	12		38.2	30.2
DEC 21	8	4	10.3	5.5
	9	3	19.8	14.0
	10	2	27.6	20.7
	11	1	32.7	25.0
	12		34.6	26.6

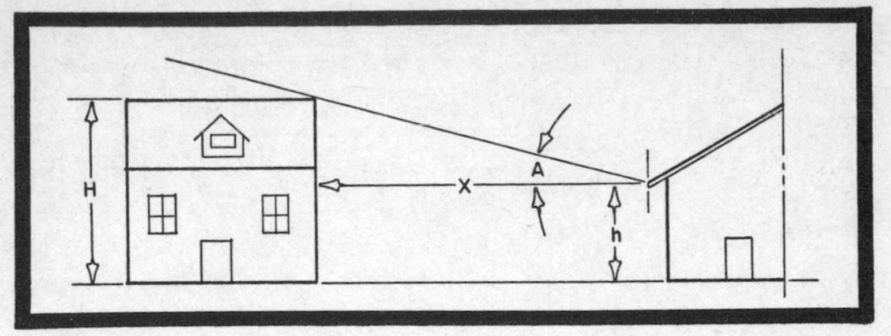

Fig. 5-9. Calculation of collector shading for various sun positions.

edge of the collectors to the top of the neighbor's house, as A. If angle A is calculated to be less than the altitude angle of the sun, this will mean the sun will be higher than the potential obstruction, and there will be no shading. Making this calculation:

$$h = 10 \text{ ft}$$
$$X = 35 \text{ ft}$$
$$H = 20 \text{ ft}$$
$$A = 16^\circ$$

$$\text{Tangent angle A} = \frac{H - h}{X} = \frac{10}{35} = 0.29$$

By referring to Table 5-2 for 40° latitude during the winter collection hours of 9:00 a.m. to 3:00 p.m., you will see the minimum altitude angle of the sun is 14° on Dec. 21—so a portion of the collector will be shaded at 9:00 a.m. But the collector will clear of shade by 10:00 a.m. when the sun's angle reaches 20.7°. Remember that all these values are based on solar time, so if you are on Daylight Savings Time, you need to adjust by one hour.

If trees are calculated to shade the collectors during the heating season and if they are *deciduous*, you can still count on some sunlight striking the collectors after the leaves fall. The bare branches, however, do cast a shadow and may intercept as much as 50 percent of the sunlight. If evergreen trees are in the way, don't count on any sunshine penetrating them.

Hail and Snow

If you live in an area that has heavy hailstorms, provisions must be made to protect the collectors from damage. Tempered glass of sufficient thickness may be used as a cover, or a wire screen mesh may be made of galvanized steel with approximately 3/8 inch square openings. Placing this mesh, commonly called *hardware cloth*, above the glass covers will afford protection, but it does cast a shadow; so if you use it, plan on a 10 percent reduction in solar insolation striking the collectors.

In the snow belt, there are some basic design features to consider. The structure must have sufficient strength to handle both the weight of the collectors and the accumulated snow. An inclination of 40° is usually the minimum angle the collectors should be tilted to prevent a heavy buildup of snow on their surface. Snow is translucent, and if there is reasonable sunshine, the collectors will warm up even if snow covers them. When this happens, unobstructed snow should slide off. Some designers use the reflection of sunlight from the snow to augment the collectors. With careful design, it is possible to use *vertical* collectors in 40° to 45° latitudes supplemented by snow reflection to obtain a net gain of energy over those inclined at the latitude plus 10° during the winter months.

Dual Glazing

You will see various recommendations on single versus dual glazing of collectors. As a general rule for the nonselective absorber designs in this book, I would suggest that you use Fig. 5-10 as a guide.

DESIGN CALCULATIONS

In Chapters 6, 7, and 8 three types of hot water preheater designs are suggested. You will frequently hear the rule of thumb that one square foot of collector is required for each gallon of hot water used per day. This is much too general and, further, is based on the assumption that it makes economic sense to collect and store water at 135° F. For most applications you end up *mixing* cold water with the supply from the water heater to cool it before use. In these days of high energy costs, tank or use temperature should be 120° F, in my opinion. The design calculations for the examples in the following chapters are based on 120° F. They offer a simplified method of calculating the required square footage of collectors, based on *averaging* conditions over the entire year rather than on month-by-month calculations. In addition, the following assumptions are made:

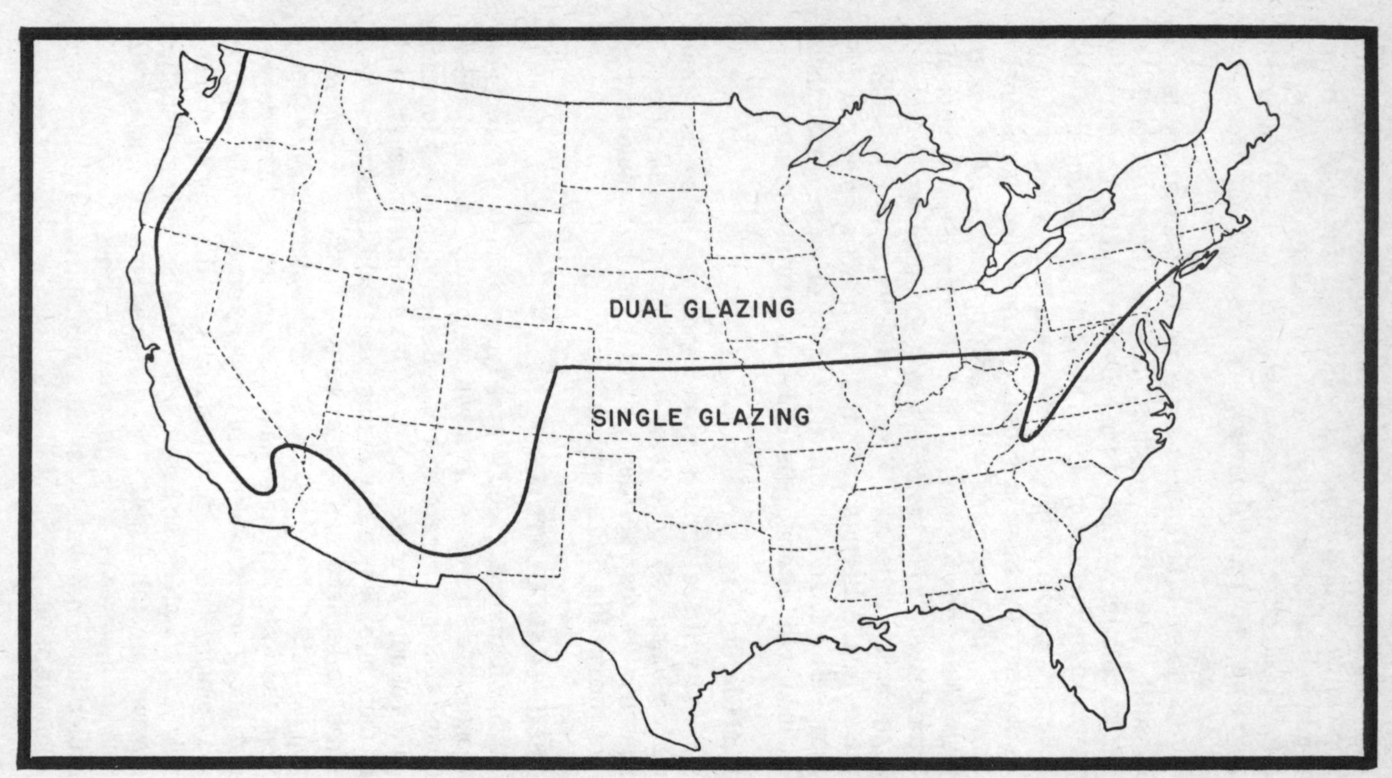

Fig. 5-10. General guidelines for single and dual glazing on flat-plate solar collectors.

- Average cold water supply temperature from city mains equal daily average outdoor temperature.
- Collectors operate at an average efficienty of 50 percent.
- Heat losses from storage tank and plumbing lines total 12 percent of the solar energy collected.
- The average family uses 20 gallons of hot water per day per person for bathing and cooking.
- An automatic dishwasher requires 15 gallons of hot water daily.
- An automatic clothes washer uses 20 gallons of hot water daily.

Again, these are *averages*, as are the *weather* conditions that solar engineers use to calculate solar input. If more precise calculations are required, I refer you to my *Homeowner's Guide to Solar Heating & Cooling*.

Here are the calculations used to develop design parameters in the example that will appear in Chapter 6:

Assumed Load:

hot water service temperature.........................120° F
daily hot water use for family of two
    with washing machine:
    personal use (20 × 2) = 40 gallon
    washing machine = 20
    total .... = 60 gallons
average daily outdoor temperature
    (annual) for Miami .... 75° F

The water heater must raise the temperature from the annual average outdoor temperature of 75° F to the service temperature of 120° F. The energy required in Btus to do this job may be calculated from the formula:

$$Q = G \times K \times \Delta T$$

where:

Q = daily energy in Btus
G = gallons used per day
K = weight of water (8.34 lb/gal)
ΔT = difference between use temperature and supply temperature

For this example:

$$Q = 60 \times 8.34 \times (120 - 75)$$
$$= 22{,}518 \text{ Btu/day}$$

Average monthly load ($31 \times 22{,}518$) ......698,058 Btus
Solar energy available (average month)
location ....Miami
latitude ....25°
collector tilt and orientation ....15° facing south

Miami has an *average* daily radiation of 451 langleys per square foot (source: *Homeowner's Guide* or U.S. Climatological Atlas). Since each langley contains 3.69 Btu, there are 1664 Btus falling on each *average* day on each square foot of *horizontal* surface in Miami. In a month, this totals (31×1664) or 51,584 Btus. But this is on a horizontal surface. If the collectors are tilted to an angle 10° less than the latitude (or 15°), they will intercept 7 percent more sunshine on an annual basis (*Homeowner's Guide* or ASHRAE) yielding a *collectible* energy of 55,195 Btus each month for each square foot of collector.

Collectible energy (Btu per month per square foot)..55,195

We assumed the collectors would operate at an *average* efficiency of 50 percent and that the line and tank losses would total 12 percent of the energy collected. The *usable* energy, therefore, is:

$$55{,}195 \times 0.50 \times 0.88 = 24{,}285 \text{ Btu per month per square foot}$$

If we divide the monthly load by the usable energy, the result is the square feet of collector surface needed:

$$\frac{698{,}058}{24{,}285} = 28.7 \text{ square feet}$$

Because each collector has ($32'' \times 74''$) or 16.4 square feet of heat-collecting surface, we can divide the square footage required by the square footage each collector provides, ($28.7/16.4 = 1.75$) to determine that two collectors are needed. This does *not* mean that supplemental energy will be unnecessary, for these are *averages*. There will be months with less sunshine and others where two collectors supply more energy than needed, with the excess being wasted. But this method will give you a good approximation without lengthy calculations.

Hot Water Preheater Using Thermosiphon Principle

The next four chapters will each describe and explain one of the four types of solar heaters. At the beginning of each of the chapters a small map will be shown highlighting the preferred climatic zone(s) for that particular system. These maps correlate with the more thorough description and explanation provided in Chapter 5 for the four individual system types. Next, the chapters will describe and provide design examples for the system being considered. It should be recognized that the systems, as defined, will not satisfy *all* load or siting conditions because these conditions vary widely.

For example, you may not have 230-volt wiring to accommodate the tank I have suggested for use with a thermosiphon system, or you may have a perfectly good gas-fired water heater and want to hook a solar storage tank in series with it. These modifications are, indeed, possible. But if you do not have a south-facing sunny area next to the building housing the water heater where *ground-mounted* collectors can be installed, then I suggest you consider a *pumped* system, as described in other chapters. My design work is intended to suggest typical systems which may be modified to suit individual requirements.

THERMOSIPHON SYSTEM DESCRIPTION

The thermosiphon principle was explained briefly in Chapter 1. Figure 6-1 shows the areas best suited for this system. No circulating pump or heat exchanger is necessary here, because tap water circulates by natural convection

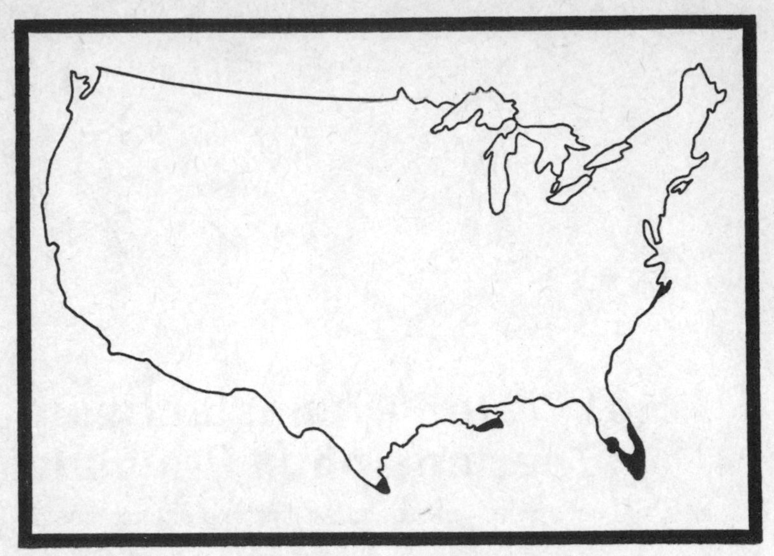

Fig. 6-1. Suggested climatic zone for thermosiphon system.

through the collectors which are ground-mounted below the storage tank (Fig. 6-2). No provision is made for automatic draining of the exposed plumbing during freezing weather; when a freeze is predicted, isolation valves must be closed and drainage valves opened manually to prevent damage from the expansion of freezing water.

The 66-gallon storage tank is equipped with a 6000-watt, 230-volt heating element which activates only when solar energy is unable to maintain the thermostatic temperature setting you select. The tank is designed for a working pressure of 150 pounds per square inch, which is greater than normal line pressure. Temperature and pressure (TP) relief valves are mounted in the top of the tank and collector return line, to provide automatic relief if excessive pressure or temperature is encountered. All plumbing, including collectors, will be operating at line pressure, and therefore requires well-made joints.

The tank design recommended is especially well suited for this system. As Fig. 6-3 shows, the collector feed is located 7 inches above the tank bottom to help prevent scale or sediment from entering the collectors. Also, the locations of cold water entry, collector feed, and collector return promote stratification zones within the tank to improve efficiency. Hot water rises to the top of the tank because it is less dense; thus,

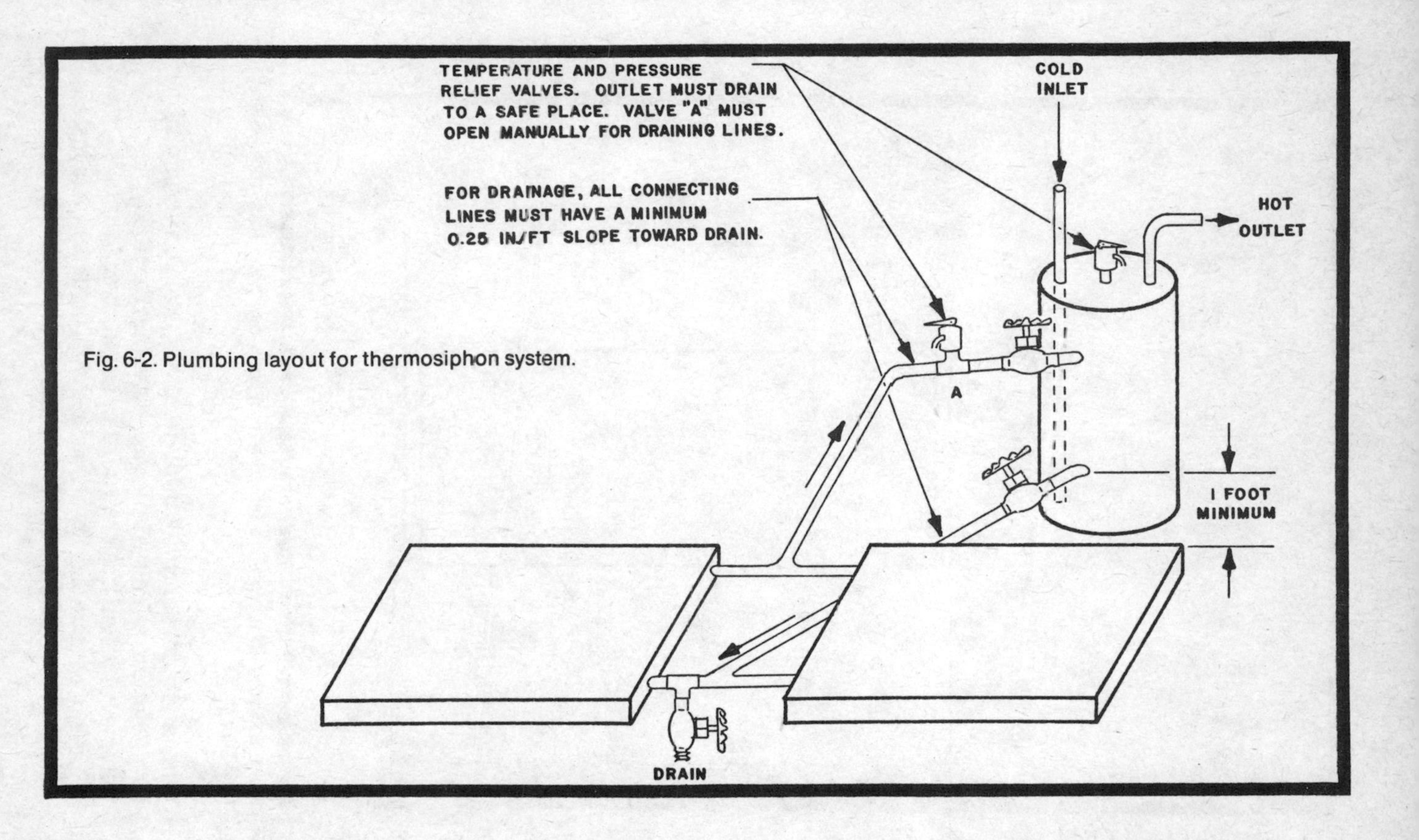

Fig. 6-2. Plumbing layout for thermosiphon system.

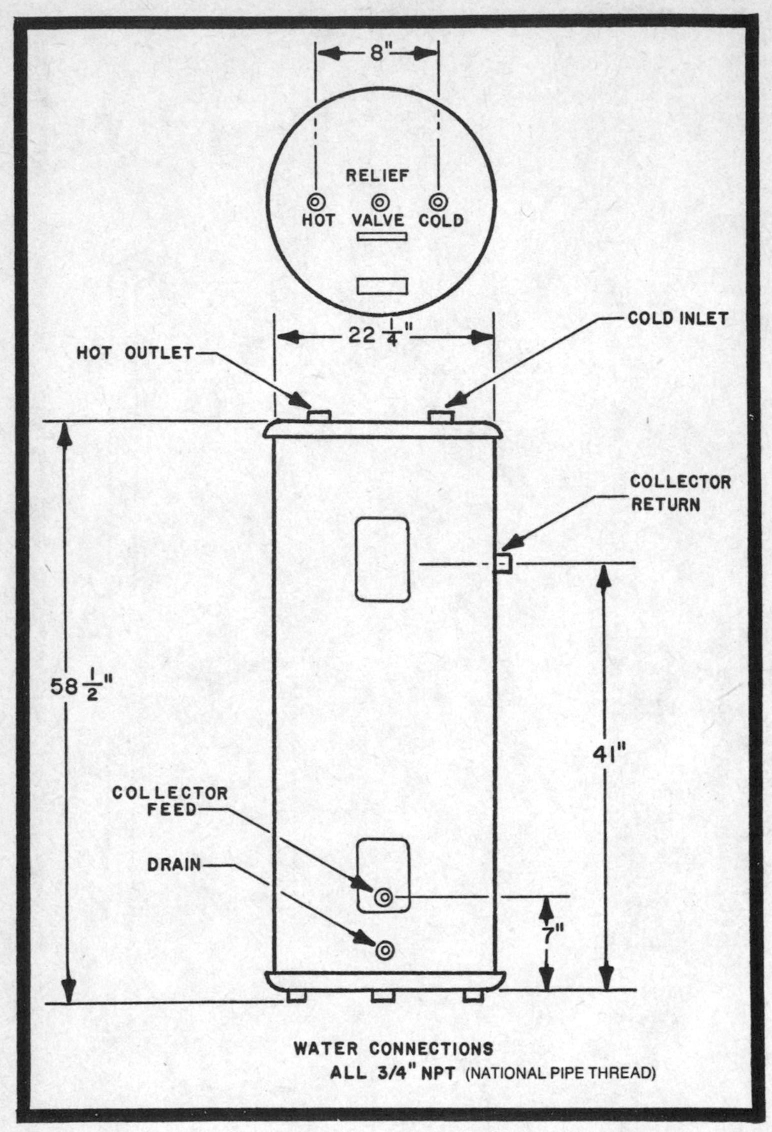

Fig. 6-3. Dimensions of Rheem 66-gallon tank designed for solar water heaters.

if agitation or *mixing* is minimized, the water will naturally *stratify* or form temperature *layers* within the tank. When hot water is required, it is drawn from the top layer.

In my opinion, the homeowner should *not* attempt to install attic- or roof-mounted storage tanks. Not only is there a potential structural problem to solve, but a tank failure or leak can cause extensive damage. Further, prevention of excessive heat losses from exposed tanks requires additional insulation. For these reasons, the system I describe should have the tank mounted on a stand near ground level.

To function properly, the top edge of the ground-mounted solar collectors must be a minimum of 1 foot, and preferably 2 feet, *below* the bottom of the storage tank. This restriction is essential to prevent the water in the system from *reversing* its flow direction at night, which would actually *cool* the storage tank.

The *rate* of flow through the collectors increases as the temperature difference across the collector increases. This means fast flow on hot days. Flow rate (at the same temperature difference) may also be increased by enlarging the vertical distance between the top of the collectors and bottom of the tank. But as flow rate increases, the temperature difference decreases, so this step can be self-defeating. Experiments have shown that a distance of 2 *feet* provides a good compromise.

To keep the collector profile low, a design was chosen that uses short risers and emphasizes the horizontal dimension. Figure 6-4 illustrates this point. In addition, a collector tilt

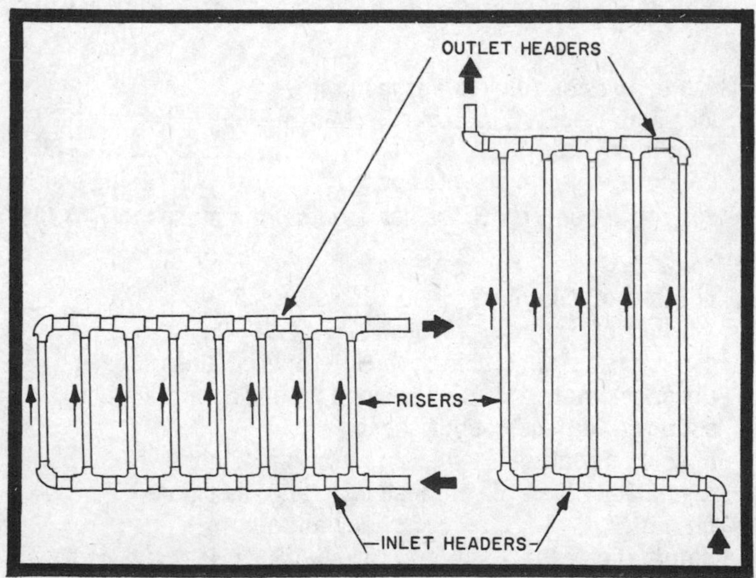

Fig. 6-4. Flow diagram and comparative height of two collector designs.

angle of latitude-*minus*-10° is suggested for units in this system, to minimize the tank height with only a slight sacrifice in performance.

The collector has an all-copper absorber to provide a fast reaction time; the copper heats more rapidly than aluminum or steel, which is especially beneficial if you have intermittent sunshine. Because each pane of glass covering the collector cuts down the amount of radiation reaching the absorber, single glazing (1 pane) is best for the geographic region specified for this system. Lastly, type M copper is recommended, to minimize cost, but if you cannot locate it, or if your local codes call for a heavier wall thickness, types K or L will do.

DESIGN EXAMPLE

Following are the calculations used to develop design parameters:

Assumed load:

hot water service temperature ....120° F
daily hot water usage for family of two with washing machine ....60 gallons
average daily outdoor temperature (annual) ....75° F
average monthly load .... 698,058 Btus

Solar energy available (average month)

location......................Miami
latitude ....25°
collector tilt and orientation ....15° facing south
collectible energy (Btus per month per square foot) 55,195

Collector design

number of collectors 2
number of covers....................1
cover material..3/16 inch thick tempered glass
absorber material ....1/2 inch type M copper risers soldered to 10-oz copper sheet
absorber coating ....flat-black paint
insulation ....industrial grade fiberglass
frame ....extruded aluminum
nominal size ....36″ × 78″ × 4″

Storage tank (Rheem 666H-66-1 or equal)

capacity ....66 gallons
heating element ....6000 watt, 230 volt
dimensions
height ....58 1/2 inches
diameter .... 22 1/4 inches
weight ....221 pounds

MATERIALS: AVAILABILITY AND COST

The materials that follow cost approximately $590 as of this writing. (Chapters 2 and 3 provided detailed information should you wish to make substitutions.) No return or feedlines are included in the estimate because the length and complexity of these lines will vary from job to job. Make your own estimate from local sources of supply, then decide on the economic advantages of proceeding with construction. The materials list that follows is based upon a two-collector system, as called for in this particular design. All subsequent lists will also be based upon two collectors, regardless of design requirements so that direct cost comparisons can be made.

Bill of Materials for a Two-Collector System

Quantity	Size	Description
Absorber:		
7 pcs.	1/2 in. diam. × 10 ft	type M copper water tube 1/2 in. (0.875 O.D.), hard temper, straight lengths
2 pcs.	1 in. diam. × 10 ft	type M copper water tube 1 in. (1.125 O.D.), hard temper, straight lengths
2 pcs.	0.021 × 30″ × 75″	10 oz (0.014 thick) copper sheet in cold rolled temper
44 pcs.	1 × 1 × 1/2	copper tees, C × C × C wrought solder-type pressure fittings
4 pcs.	1 × 1/2	copper 90° ells, C × C short radius, wrought solder-type pressure fittings

Quantity	Size	Description
Collector frame:		
4 pcs.	10 ft	aluminum frame extrusion, Alcoa die number 459831
4 pcs.	10 ft.	aluminum glazing cap extrusion, Alcoa die number 459841.
1 pc.	2 ft	aluminum hold-down clip extrusion, Alcoa die number 459851
60 pcs.	No. 10 × 3/4 in.	No. 10 type A, round or pan head sheet-metal screws, 3/4 in. long, cadmium plated steel
Back plate:		
2 pcs.	1/4″ × 4′ × 8′	exterior grade or marine plywood, 1/4 in. thick
Collector insulation:		
14 pcs.	1″ × 24″ × 48″	industrial grade semi-rigid fiberglass board (Certain-Teed Corp. No. 850, Johns-Manville No. 814 or equal), no facings
Glazing:		
2 pcs.	3/16″ × 34″ × 76″	tempered glass with low iron content
1 roll	1/4″ × 1/4″ × 50′.	General Electric Silglaze tape SCT 1534 or equal
1 cartridge	1/12 gal	General Electric Silglaze silicone sealant cartridge SCS 1803 or equal
1 cartridge	1/12 gal	General Electric 1200 series silicone sealant cartridge SCS 1203 or equal
1 doz	1/8″ × 1/8″ × 1/2″.	hard neoprene rubber setting blocks

Quantity	Size	Description
Valves:		
2	3/4″	isolation valves—may be either gate or ball type. Cast brass with 3/4 in. NPT threads both ends
1	3/4″ × 3/4″	boiler drain valve. 3/4 in. male NPT thread on inlet and 3/4 in. threaded hose connection on outlet. Cast brass construction
1	3/4″ (4″ long element)	temperature and pressure relief valve. 150 psi, 210°F type NHL TP valve.
1	3/4″ (short element)	temperature and pressure relief valve. 150 psi, 210° F type NHL TP valve.
Miscellaneous plumbing:		
2	1″ × 3/4″	copper adapters. C × M wrought solder-type pressure fittings. 1 in. side is female solder-type, 3/4 in. side is male 3/4 in. NPT thread
2	1″ × 1″ × 3/4″	copper tees, C × C × C wrought solder-type pressure fittings
2	1″ × 1″ × 1″	copper tees, C × C × C wrought solder-type pressure fittings
4	1″ × 1″	copper unions, C × C wrought solder-type pressure fittings
1	1″	CPVC coupling
1	1″ × 3/4″	copper adapter. F × M NPT threads both ends
1	1″	copper pipe cap. NPT pipe threads

Quantity	Size	Description
Storage tank:		
1	66 gal.	Rheem model 666H-66-1 or equal
General items:		
1 qt		heat-resistant flat-black paint
3 lbs	1/8 in. diam.	50A wire solder (50% tin, 50% lead by weight)
1/2 lb		good quality paste flux. LA-CO (Lake Chemical Co.) or equal
as needed	1″ diam.	type M copper water tube 1 in. (1.125 O.D.) hard temper, straight lengths (to connect collectors to storage tank)
as needed	1″ I.D.	closed-cell foamed rubber tubing insulation. 1 in. minimum wall thickness (to insulate connecting lines)
as needed		adhesive to cement foamed rubber tubing insulation

In the sections that follow the descriptions relate to the fabrication and assembly of *one* collector only. If you need more than one, additional pieces should be cut as required.

FRAME FABRICATION

The collector frame is made from standard aluminum extrusions available from Alcoa. I suggest that before you begin the cuts you refer to the section on collector frame fabrication in Chapter 1. A cross section of the parts, which includes a frame, glazing cap, and hold-down clip, is shown in Fig. 1-14. Three kinds of saw cuts are required on these parts: a 45° bevel and 90° rip on the frame, 45° bevel on the cap, and a straight 90° cut on the clip. I found a radial arm saw to be the easiest to use, and the description that follows gives you my method. Remember to wear safety goggles while sawing.

For this particular collector, which is single-glazed (one cover), you will need to saw off the upper leg of the frame extrusion (this leg is provided for dual glazing only). The length of the leg after sawing should be the glass thickness (3/16 inch), plus an allowance for the setting block (1/8 inch), plus an allowance for the glazing tape *after* compression (1/8 inch)—or a total of 7/16 inches. (If you look ahead a little, this distance is shown as dimension L in Fig. 6-10.) Position the 10-foot long frame extension on the saw table with the legs up and the saw in the rip position. Cut the leg to the required dimension on two of the 10-foot frame pieces. With the saw blade perpendicular to the part, tilt the blade to a 45° *bevel* angle. Now position one 10-foot long frame extrusion on the saw table with the legs *down* and a 10-foot length of glazing cap resting on top of it. The parts should be in assembly position, as was shown in Fig. 1-17. By cutting both pieces together you can save time and insure that the length of each cap corresponds to its mating frame member.

Measure a distance of 77 1/4 inches on the wide, flat surface of the frame and proceed to cut both pieces to length. Figure 6-5 shows the setup. The pieces must now be beveled on their other end, so turn the assembly end for end and make a

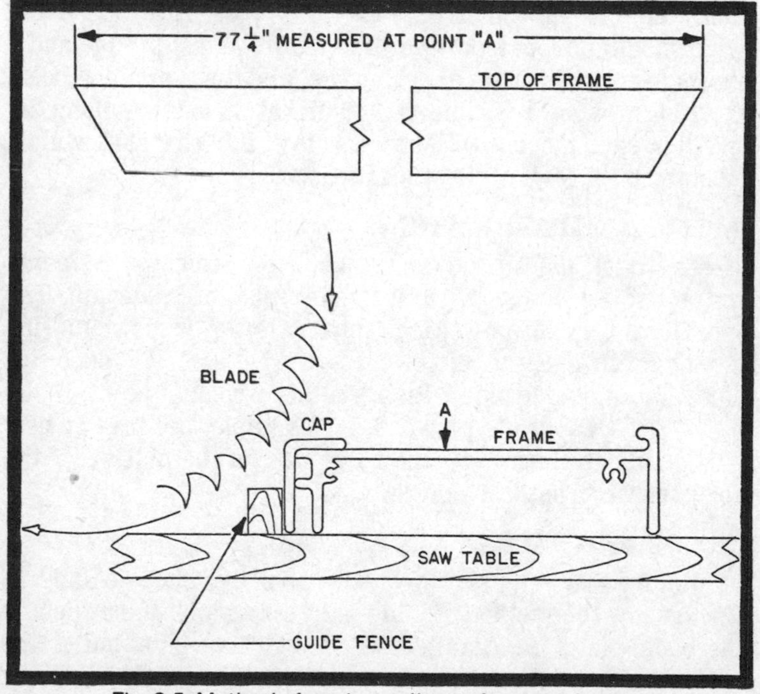

Fig. 6-5. Method of sawing collector frame and cap.

similar bevel. Be sure to hold the pieces tightly together so they will not shift. When making this cut, *do not* shorten the pieces. The frame should still measure 77 1/4 inches after beveling both ends. Repeat this step for the other side member of the frame and the two frame ends. The end pieces should measure 35 3/8 inches after beveling. In each case, cut the cap along with the frame.

Drilling is the next step, and if you refer to Fig. 1-14 again you will see that there are locator grooves in the side of the frame extrusion which identify the center line of the screw slots. It will be necessary to drill four holes into each *end* piece of the frame for assembly. Measure 1/4 inch in from each end of the 35 3/8 inch long frame pieces, along the screw slot locator grooves. After a light center punch, drill four 5/64 inch diameter *pilot* holes in both parts. Now use a 3/16 inch diameter drill to enlarge the holes. This may sound like an extra operation, but a large drill will tend to wander without the pilot hole and the 1/4 inch edge distance is *critical* to proper frame assembly.

Use the same procedure to drill attachment holes on the back flange of all four frame extrusion pieces. The 3/16-inch holes should be located 3/8 inch in from the inside edge and 2 inches in from the ends of each piece, and the remaining holes should be spaced approximately 8 inches on centers along the length of each piece. In a later operation, the back plate will be mounted by screws at these locations.

BACK PLATE FABRICATION

The back plate protects the collector from the elements and must be made of a weather-resistant material. The specifications call for 1/4 inch thick *exterior*-grade or marine plywood. The piece should be cut to a 35 1/4-inch × 77-inch size and painted on one side with a weather-resistant paint. If you use a "B – C" grade plywood, for example, the "B" or best face should be painted; this side goes on the outside of the collector when assembled.

FRAME AND BACK PLATE ASSEMBLY

Before starting assembly, refer back to Figs. 1-15 and 1-16 which show the method of joining the extrusions and attaching the back plate. As detailed, the back plate slips under the bottom screw slot on the frame extrusion and is held snugly by screws.

Clean the mitered corners and the recess under the bottom screw slot on the four frame members by wiping these areas with lacquer thinner or MEK (methylethylketone). Apply a bead of silicone sealer (Silglaze or equal) under the screw slot and along one mitered edge of a frame side and end piece. Assemble these to the back plate and join the mitered corner by using two number 10, type A, pan head, sheet-metal screws, 3/4 inch long. These should be cadmium-plated steel or stainless steel. These screws will be difficult to tighten as they thread themselves into the screw slots. If you find they drive easily, you probably missed the slot or are using the wrong size screw. Look inside the frame to be sure the screw is engaging. Once the screws are tight, caulk the corner from the inside with silicone sealer (Silglaze or equal) until the material extrudes through the joint to the outside edge. Wipe off the excess on the outside. Repeat this procedure until all four frame parts are secured at the corners.

Now turn the frame assembly upside down with the plywood on top and screw the frame to the back plate. You can use the same number 10 pan head screws for this operation. It helps to drill a small 5/64-inch pilot hole into the wood first to start the screw. It takes approximately 24 hours for the silicone to fully harden, so this assembly should be set aside and allowed to dry.

ABSORBER FABRICATION

The tubular portion of the absorber assembly consists of 1/2-inch (nominal size) type M copper water tubing used as *risers* connected by means of solder-type tee pressure fittings to 1-inch, type M copper *headers*. The risers are spaced 6 inches on centers and they have 10-ounce (0.014 inch thick) copper sheet soldered to them. To understand what is meant by riser and header, refer back to Fig. 6-4. When the collector is operating, water flows into the *inlet header* which is in a horizontal position. When it is filled, the water begins to flow upward through each of the *risers* and is discharged through the *outlet header*. Figure 6-6 shows the details of how the assembly between riser and header is made.

Begin by taking three 10-foot lengths of 1/2 inch tube and cutting each into four equal pieces, yielding a total of 12 pieces, 30 inches long. Use a tubing cutter to do this, as described in Chapter 1 (Fig. 1-5). These will be used to make the risers for the collector.

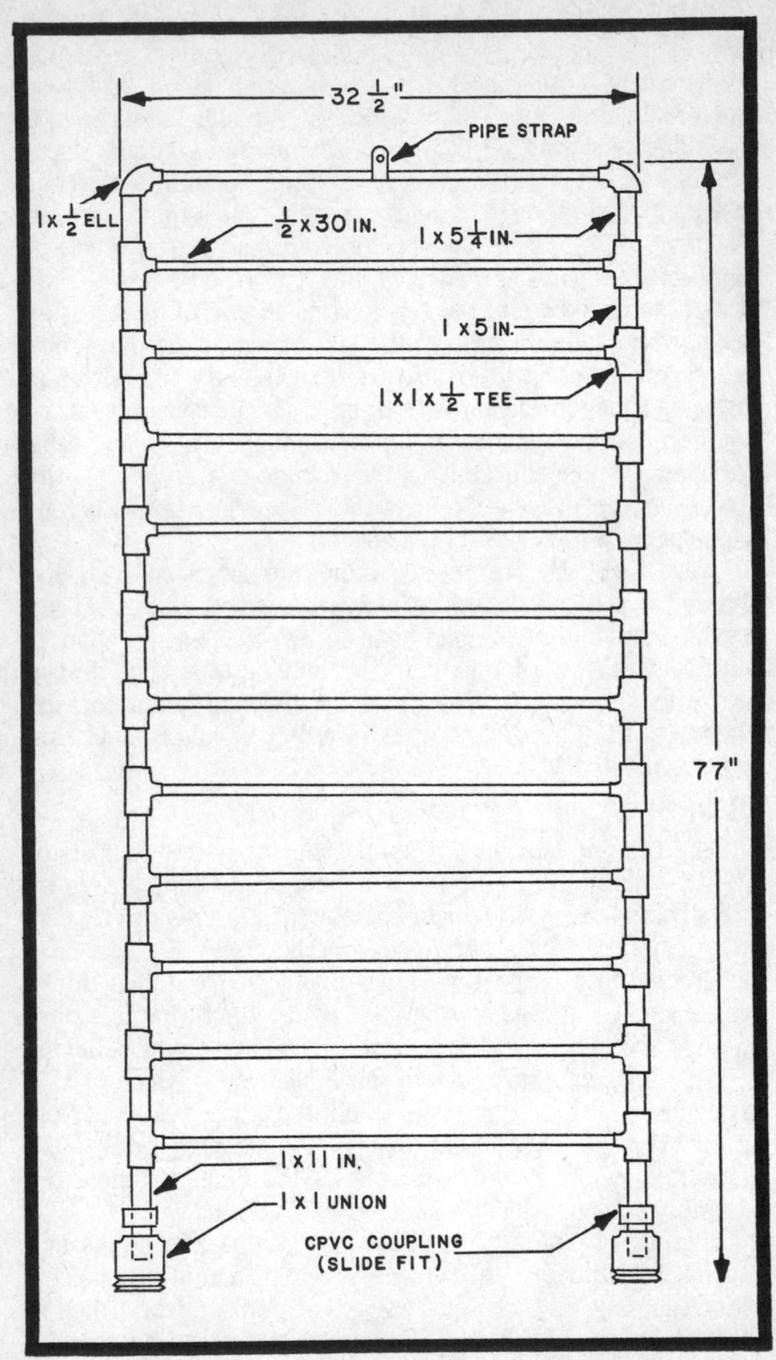

Fig. 6-6. Assembly diagram of absorber components.

Next, cut the 1-inch tube into: 20 pieces, 5 inches long; 2 pieces, 5 1/4 inches; and 2 pieces, 11 inches. These will be assembled between the tees, ells and tees, and tees and unions, respectively. If you saw the 1-inch tube, be sure to file off the burrs. And regardless of the cutting method, ream all pieces on the inside using a tube reamer, as pictured in Fig. 1-6, or a half-round file.

Soldering the Absorber Plate

Lay out the pieces you have cut, along with the tees and ells, positioning them as shown in Fig. 6-6. Clean and flux all joints, then insert the tubes in their respective fittings, making sure they go in all the way and bottom out. When the parts are in alignment, place the assembly on a flat work surface, extending one of the headers over the edge by about 8 inches. Again check the alignment, then begin to solder the header to the risers. Use the method outlined in Chapter 1, starting at the *center* riser and working out toward each end, soldering one tee at time. Once the assembly has cooled, turn it end for end and solder the remaining header. Be sure the two headers are in alignment and not twisted relative to each others. Solder the copper pipe strap in position. This is used to anchor the absorber in position, as detailed in Fig. 6-7.

The 1-inch CPVC coupling should be cut into two pieces, thereby removing the *ridge* in the center.Ream the inside of the pieces as necessary until they can be slipped by hand along the 1-inch copper tube at the ends of the headers. Slip one piece of plastic on each tube and solder the *male* portion of the couplings, called *unions*, to the open ends of the inlet and outlet headers.

The next operation is to solder the 10-ounce copper sheet to the riser tubes. The 30″ × 75″ sheet is placed on a flat surface and centered so it just touches the headers and has an equal overhang on each side of the risers. Trace around the tubes with an awl to mark their location on the sheet. Remove the tube assembly and clean and flux the sheet and tubes where the solder joints are to be made. Reposition the assembly in preparation for soldering.

The technique I used was to place a 1 × 2, 30 inches long, *under* the center rise, then to stand on a 2 × 4, perpendicular, on top of the riser as necessary to keep the sheet and tube in close contact while soldering. Starting with the center tube and

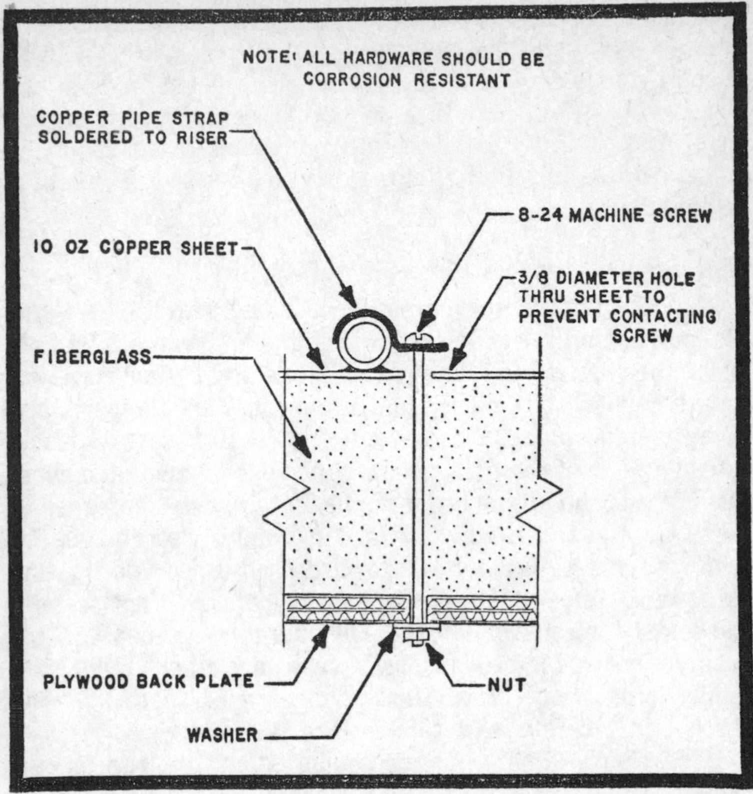

Fig. 6-7. Suggested detail for anchoring the absorber plate.

working toward the edges allows for some expansion and minimizes buckles and waves in the sheet.

There are a few soldering tips that will be helpful in making this assembly. You should start by playing the flame alternately on the tube and sheet, keeping the flame away from the fluxed and cleaned areas to avoid oxidation. The thin sheet heats rapidly, so after initial warming, keep the blue inner cone of the flame right on the *tube* until the copper begins to lighten in color. While the tube is heating, dip the wire solder into the paste flux. Apply the solder to the fillet between tube and sheet, and as it melts, move the flame and solder together *slowly* along the length of the tube, continually fluxing the solder as it is used. Use the board to keep close contact between the parts when necessary, because the sheet tends to buckle as it is being heated. To gauge your time, this assembly takes about 4 hours.

Pressure Testing the Absorber

The absorber assembly must be pressure tested for leaks before it is mounted in the collector frame. If there is a problem with the soldered joints you have made, now is the time to repair them. In the bill of materials, I listed a 1 inch × 3/4 inch F × M adapter and a 1-inch cap. These fittings have national pipe threads (NPT) and are used to make the testing fixture. Assemble the fittings to the male unions and, using a section of garden or appliance hose with 3/4 inch female fittings on both ends, connect the absorber assembly to a water line. With the cap fitting removed or loose, turn the water on slowly to *bleed* the air out of the absorber; then tighten the cap and pressurize the assembly. Wipe the absorber completely dry, then watch for any water leaks. I would suggest that you leave the pressure on for several hours. If leaks do occur, resolder the faulty joints.

Painting the Absorber

The top of the absorber plate should now be cleaned and painted before final assembly. Paste flux may be removed from the copper with an organic solvent such as lacquer thinner or MEK (methylethylketone). After the flux has been removed, scrub the metal with an abrasive household cleaner, such as Ajax or Comet, until the surface is relatively shiny; then rinse. You now have the choice of either using a heat-resistant flat-black paint available at hardware stores—of the type used for painting barbecues—or going to the more expensive industrial grades that are designed for solar applications and have excellent durability. The barbecue paint requires no primer, comes in an aerosol spray can, and is easy to apply. The disadvantage is that you may get some peeling as the absorber plate expands and contracts in use. If so, it would require removing the glass cover from the collectors and repairing the paint. There are two good industrial products on the market: Nextel 101-C10 Black Velvet coating, manufactured by the 3M Company, and Enersorb flat-black base and activator (two-part coating), manufactured by DeSoto, Inc. Both products require a separate primer and may be applied with a brush, but spraying is recommended. If you are interested in these products, you should consult the manufacturers' literature for specific application instructions.

MOUNTING AND INSULATING THE ABSORBER

The collector must be insulated around the inside perimeter of the frame and behind the absorber. I chose 1 inch thick, industrial-grade fiberglass board for the frame and the same material, 2 inches thick, behind the absorber. The material specified has high temperature resistance and will withstand the static (non-flow) temperatures that can build up inside the collector. In addition, it has a 3-pound density which should be sufficient to support the weight of the absorber. Substitutions may require other mounting methods.

Cut one piece of the 1 inch thick fiberglass board into 3 3/8 inch wide strips 48 inches long. You will need 5 strips which will be used to line the inside perimeter of the extruded frame. Place the strips along two sides and one end of the frame, tucking the top edge under the flange of the extrusion and pressing the bottom in place. Cut 6 pieces of 1 inch thick fiberglass 24 inches wide by 33 1/2 inches long, using the remainder of the material from the 24″ × 48″ sheets to make two pieces 3 inches by 33 1/2 inches. When assembled, these sections will make an insulation blanket 2 inches thick × 33 1/2 inches wide × 75 inches long which goes between the absorber and the back plate. Lay these pieces in position, completing the insulation except for one end of the frame.

Lay the absorber on top of the insulation with the tubes on top. It may be necessary to bend up one edge of the copper sheet, but this can be flattened later. With the unions pressed against the uninsulated frame end, press down on the absorber, keeping it tight against the insulation, and mark the position of the unions on the extrusion. Remove the absorber and back insulation. Using a coping saw or metal hole saw, cut two holes through the frame at the location marked. The diameter should be just large enough to allow the unions to slip through. Reassemble the insulation including the end piece which whould be pierced to allow for the unions.

Install the absorber, centering it in the collector box. To hold the end opposite the unions tightly in place, an anchor like the one detailed in Fig. 6-7 is used. Tighten the screw until the risers are an equal distance below the top of the frame on both ends of the absorber. Make sure the 1-inch CPVC insulating collars are *between* the headers and frame, then caulk around the CPVC with a bead of silicone sealer.

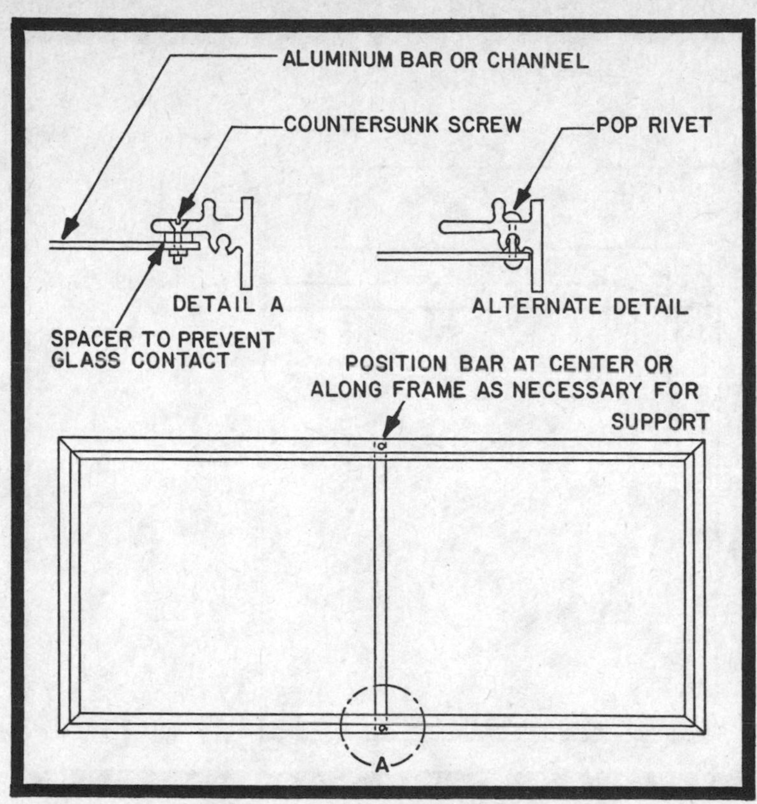

Fig. 6-8. Suggested details for mid-span support of collector frame.

The last operation before glazing is to support the *center* of the frame side members to prevent them from bowing outward and losing contact with the glass. A small aluminum bar is used to tie the sides together. It is held in position by pop rivets or sheet-metal screws as shown in Fig. 6-8. Measure the glass opening to insure that the dimension is correct and constant.

GLAZING THE COLLECTOR

To prepare for attachment of the glazing caps, drill a 3/16 inch diameter hole 2 inches in from the ends of each cap piece along the locator groove. Space the remaining 3/16 inch holes approximately 8 inches on centers along the length of each piece. Place the glazing caps on top of the corresponding frame members, marking the location of the drilled holes on the top leg of the frame pieces. Remove the caps and drill

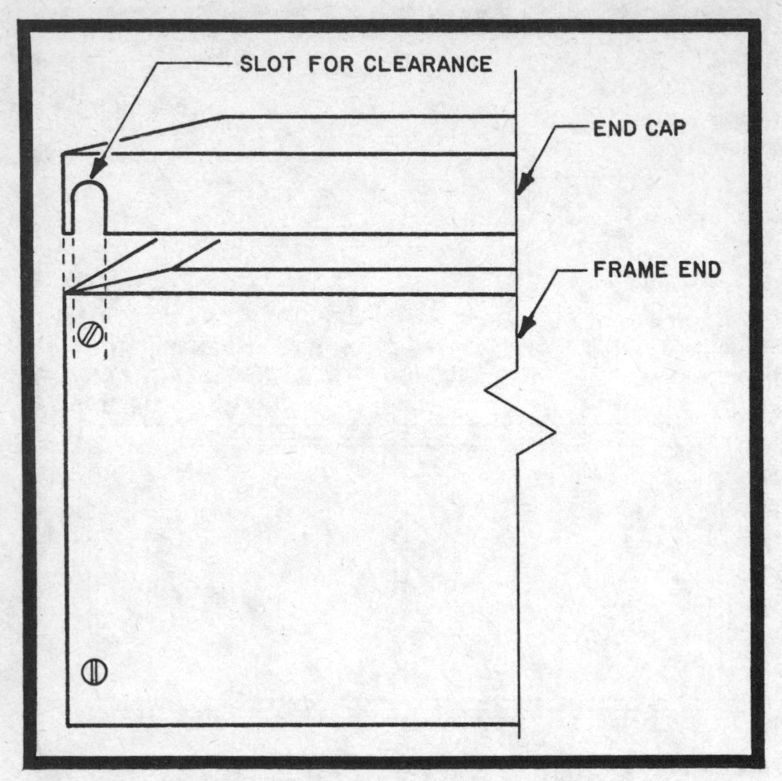

Fig. 6-9. Slot for clearance of glazing cap.

5/64-inch diameter pilot holes at the locations marked. Now enlarge the pilot holes with a 1/8 inch drill. Don't substitute another drill size, as a number 10 sheet-metal screw must thread itself into these holes.

Lastly, there must be clearance between the glazing caps on the frame ends and the *top* screw holding the frame together. With the cap held in place, mark the location of the screw slot hole from inside the frame. Drill a 1/4-inch hole through the cap, then enlarge the hole to a 3/8-inch slot with a round file. This will allow the cap to slip over the screw during assembly. Figure 6-9 shows this detail.

When you are ready to glaze, refer to Fig. 6-10. Lay the collector flat and clean the surfaces of the aluminum frame and glazing caps that will be in contact with sealant or tape. Use methylethylketone cleaner, commonly called MEK by the trade. Also wipe the inside and outside surfaces of the glass cover in a 1-to-2-inch strip with the cleaner. The glazing tape comes in a roll with a release paper on one side and a pressure-sensitive adhesive on the other. Unroll the tape, leaving the release paper on it but applying the adhesive side to the vertical glass stop around the perimeter of the frame. This acts as a resilient spacer, protecting the glass edge from the metal yet allowing for thermal expansion. Place the 1/8 inch thick setting blocks at quarter points around the frame using a small amount of silicone to hold them in position.

Now apply a continuous bead of *high temperature* silicone along the inside edge of the horizontal leg holding the glass. A silicone rubber sealant system capable of withstanding 325° F is preferred. You may not reach that temperature with a single glazed unit, but the extra precaution is worth the investment. I chose General Electric Silicone Construction Sealent SCS-1200 system. The silicone bead should be approximately 1/4 inch in diameter and should extend above the glass stop to insure contact with the glass. Being careful not to contaminate the glass edges just cleaned, gently lay the glass into position. I would suggest wearing clean white cotton gloves when handling the glass.

Apply glazing tape to the glazing caps, locating it 1/4 inch in from the edge. Unroll the tape, leaving the release paper on it but applying the adhesive side to the underside of the glazing caps. Use a razor blade to cut the tape to length corresponding to the 45° miter on the parts. Remove the release paper from

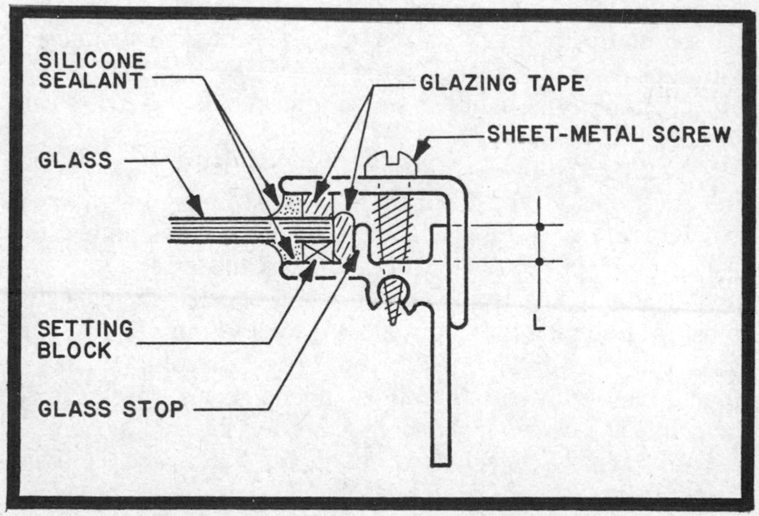

Fig. 6-10. Typical detail for single-glazed collectors.

the tape on the glazing caps and apply a bead of sealer next to the inside edge of the tape using the caulking gun. This bead should be large enough to extend to the top edge of the tape or about 1/4 inch in diameter. Put the four glazing caps in place, one at a time, tightening the retaining screws by hand. Apply sealer from the gun along the mitered corners as the pieces are assembled.

Systematically, and evenly, tighten the retaining screws along the glazing caps until the inside edge of the cap extrusion is a minimum 1/8 inch and maximum 3/16 inch from the glass. This will mean you have compressed the outer 1/4 inch glazing tape by 25 percent to 50 percent, which is generally recommended. But in all cases throughout this book, the specifications of the individual manufacturers should be followed for the products used.

TANK LOCATION AND INSTALLATION

As mentioned previously, it is essential that the hot water tank be elevated above the collectors by a 1-foot minimum for thermosiphon action to occur. If your home has a crawl space below the floor, this should not be a problem; the collectors have a low profile and in southern latitudes should not extend more than 15 inches above ground level. The vertical distance is actually measured from the collector feed line on the tank (cold water outlet of tank) to the collector outlet header. For homes with floor slab construction, it may be necessary to build a platform to elevate the tank the necessary distance. If the tank is not in the living space, this should not be objectionable; but in any case, your local building code should be checked.

Another requirement governing the tank location is that the feed lines to and from the collectors be kept as short as possible and be well insulated. I would not recommend that you go much beyond 10 feet of piping on each leg or a total of 20 feet of pipe to and from the collectors. Because the collectors must face south and be unshaded from 9:00 a.m. to 4:00 p.m., your options will be limited. If a suitable location cannot be found, you may be forced to abandon the thermosiphon principle and consider the more expensive pumped system.

The tank I selected is called Solaraide and is manufactured by Rheem Water Heater Division. The product is available in six models: three with a 6000-watt electrical heating element and three without any supplementary heating source (storage only). Capacities of 66, 82, and 120 gallons are available.

Plumbing Connections

Even though codes may require a pressure-only relief valve installed in the water tank, I suggest that you use a combination temperature and pressure (TP) unit certified to ANSI Z21.22 standards. These are set to automatically open if the tank pressure exceeds 150 psi or if the tank temperature exceeds 210° F. (Check the manufacturer's specifications, because these values vary somewhat). It is possible for these upper limits to be attained if no water is used during hot, summer weather (if you go on vacation, for example); in this case, the valve opens automatically, discharges, and reseats. This value is simply screwed into the top of the tank in the opening marked *relief valve*. Figure 1-22 showed a typical valve. As stated earlier, a drain line must be attached to the valve and must terminate in a safe place to prevent scalding in case of discharge.

Two isolation or shut-off valves are placed where the collector lines join the tank. These are made of cast brass and are of the gate or ball type. Do not use a globe valve, because flow is more restricted through this type of valve and the globes can impede thermosiphoning. Gate valves have a *port size* (an opening inside the valve) equal to the mating pipe for full flow. A ball valve opens and closes with 1/4 turn of an operating level and does not provide quite the flow efficiency of a gate valve. For this application, the isolation valves will be operated fully open when the collectors are used, or fully closed when they are being drained. The valves will not be used for throttling. Regardless of your choice, buy the isolation valves with a 3/4-inch national pipe thread on both ends, screwing them onto the galvanized steel nipples provided on the tank for the collector return and feed lines. Use C × M brass adapters to connect the 1 inch diameter copper feed and return lines joining the tank and collectors. A C × M fitting has male threads on one end (in this case, 3/4 inch size) and a solder-type joint on the other (1 inch size).

A drain valve is installed at the lowest point in the plumbing; that is, where the inlet headers on each collector join the cold water feed line. I used a 3/4-inch faucet with a hose connection called a boiler drain. The TP valve installed outside the isolation valve on the collector return line functions both as a safety device and an air relief valve. When its test lever is opened, air enters the line to facilitate draining the collectors. Keep in mind that the five valves used must be rated to handle the line pressure of 125 pounds.

This design used 1-inch diameter copper lines to connect the collectors to the tank. The 3/4-inch lines may be suitable if each line does not exceed 10 feet in length. Keep the lines as short as possible with a minimum of obstructions (bends, fittings, etc.). Ideally, the lines should be sloped with a uniform rise from collector to tank; but realistically, some sections will be horizontal or vertical. The important point to remember is to avoid any sections that fall *below* the horizontal, thereby causing air pockets. For example, looking at Fig. 6-11, if air did get into the line, it will *rise* to the highest point but will not travel downhill. The design on the left would cause air entrapment; the one on the right allows air to bubble up to the top of the storage tank where it does not restrict flow and is eventually discharged through the house faucets.

Soldering the headers from the collectors to the feed and return lines is straightforward, using techniques previously described. But before making any soldered joints in the plumbing, assemble all the pieces as shown in Fig. 6-2,

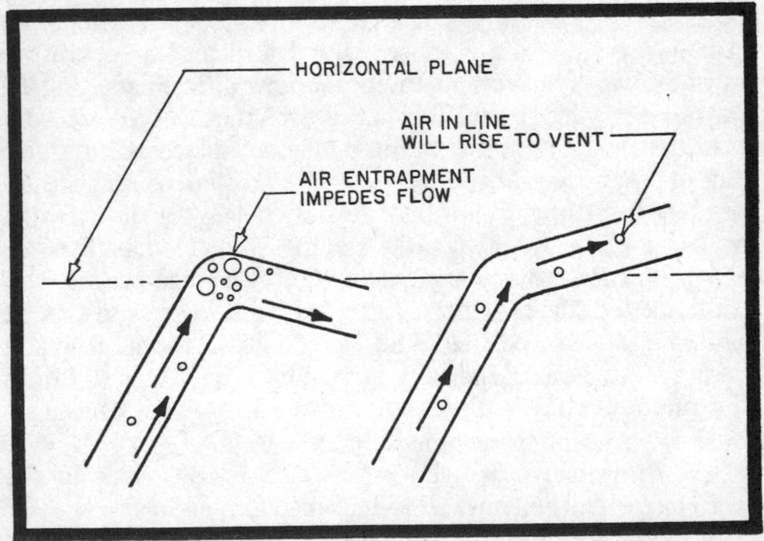

Fig. 6-11. Effect of plumbing line slope on air entrapment.

cleaning and fluxing the parts first. Once everything is in alignment, solder the parts together.

The next step is to pressure test the plumbing before final sealing and insulating operations. Because the absorber assembly was previously checked, no leaks would be expected *inside* the collector(s). Make sure that the isolation valves are open and drain valves are closed (both on the collector *and* the tank). To fill the tank with water, open the shut-off valve in the cold water supply line, and while the tank is filling, open a hot water faucet in the highest location in the house. Keep it open until water flows from it in a steady stream. Now close the faucet and carefully inspect all the plumbing lines you have installed. If leaks are present, you must drain the lines that need repair and resolder them.

Sealing and Insulation

Check the openings between the inlet and outlet headers and the extruded frame again, making sure they are sealed against the weather.

The feed and return lines between the collectors and tank must be well insulated to prevent heat losses. I would suggest that the material used have a minimum R-4 rating and be water resistant.

A 1-inch thick layer of closed-cell urethane tubular pipe jacketing with a 1 inch I.D. should meet both requirements. Because this material is applied after joint assembly and testing, it cannot be slipped over the pipes. It must be slit uniformly along one side once the pieces have been cut to the proper length. Make cutouts for fittings and valves as necessary, being sure to cover them with insulation. A sharp knife will cut the material, but a serrated blade such as a bread knife will give better results. After the insulation has been cut to size, all joints and slits in the lines must be sealed using adhesive recommended by the manufacturer. The insulation may be painted with exterior latex or vinyl base paint to improve its appearance and weatherability.

Drain Holes

A provision should be made for drainage in the event that water penetrates the solar collectors. Holes drilled through the collector frame on the *low* side are suggested. These may be referred to as drain or *weep* holes. Drill two 1/4-inch diameter noles along the screw slot locator groove at the bottom of the lowest frame member. This will penetrate the collector just above the back plate, allowing drainage if necessary.

OPERATING THE SYSTEM

You must consult your local electric company before connecting the heater, to insure that present wiring is of sufficient size to handle the wattage load for the heater specified. All wiring must conform to the local code or National Electric Code.

Solaraide heaters come completely wired to the junction bracket on top of the tank. Removing the cover plate, you can see one red and one black wire plus a grounding screw. In Chapter 1, I described 120-volt electrical connections using black and white color-coded wiring. For heavy appliances, such as an electric water heater operating at 240 volts, a white or grounded neutral wire is not used; in this case the wire is red. If you currently have a 250-volt outlet rated for 25 amps or more, you may be able to connect the heater yourself with some suggestions from the electric company representative. If not, he will probably suggest that a qualified electrician be employed.

Before connecting the power, remove the cover plate on the heater with the Solaraide decal. Push back the insulation to reveal the thermostat adjusting screw. Turn the screw until the desired water temperature is shown on the indicator. If you are an energy-conscious individual, I suggest the setting be 120° F. Before replacing the insulation and cover plate, observe the small red button on the thermostat. If for any reason the water temperature in the tank becomes excessively high, the *high limit control* breaks the electrical circuit to the heating element, and this red *reset* must be pressed to reconnect the circuit. This must not be done, however, until the *cause* of the high temperature is located and corrected.

I advise you to switch off the power to the water heater before leaving on a summer vacation. If hot weather prevails and no water is used, the solar collectors may provide enough heat to trip the high limit control.

As mentioned earlier, the collector feed line is raised 7 inches above the tank bottom to help prevent scale or sediment from entering the solar collector system. To maintain efficiency, it is recommended that a few quarts of water be drained from the tank every few weeks to flush lime and sediment deposits from the bottom of the tank. Draining procedure should be done in accordance with the manufacturer's instructions.

Maintenance of the system is relatively simple. Wash the dust and dirt from the glass covers periodically. If you live in an area where the water has a high mineral content, scale will tend to build up in the absorber tubes over a period of years. This will restrict the water flow and decrease collector efficiency. If you have such water conditions, plan to flush the system as required, using commercially available solvents approved for potable water systems.

If you want to minimize the amount of auxiliary energy (electricity) required, wash dishes and laundry early in the morning. The temperature in the tank can then be restored during the day, with the sun as the primary energy source, in preparation for the evening demand for hot water.

7 Hot Water Preheater With Circulating Pump and Automatic Drain-Down

Most homeowners are reluctant to monitor weather conditions in order to manually drain their systems in advance of freezing weather—the practice required when using the system described in the last chapter. It is for this reason that I restricted the thermosiphon system to the small geographic region A shown in Fig. 5-2 or the shaded areas in Fig. 6-1.

Providing freeze protection by means of an automatic *drain-down* feature allows the system in *this* chapter to be recommended for either the B or C regions of Fig. 5-2, or those areas shaded in Fig. 7-1. Dual glazing is not normally recommended.

SYSTEM DESCRIPTION

A circulating pump is used to force potable water through the collectors at a controlled rate in the hot water preheater system with a circulating pump and automatic drain-down. This feature allows the collectors to be mounted above the storage tank, which is the most popular location. The use of tap water through the collectors, coupled with the elimination of an intermediate heat exchanger, provides good system efficiency. As in the Chapter 6 system design, the collectors for this system, and all plumbing connections to it, must be constructed to withstand line pressure up to 100 pounds per square inch.

Safety features are built into the system to prevent excessive pressure or temperatures. Pressure-relief valves are installed on the storage tank and in the line leaving the collectors. These valves are designed to open automatically at

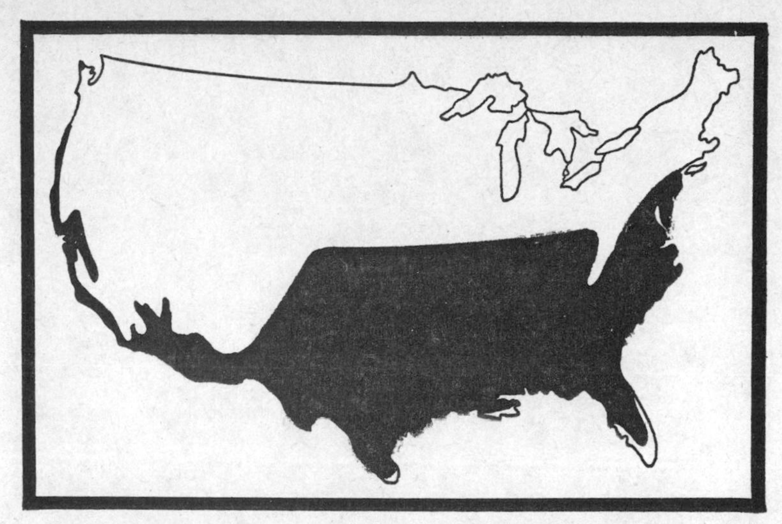

Fig. 7-1. Suggested climatic zone for automatic drain-down system.

100 psi pressure, to discharge, and then to close once the excess pressure had been relieved. The differential thermostat is equipped with a high temperature control which automatically turns the circulation pump off when a storage tank temperature of 140° F has been reached. Solar collection in excess of this temperature would cause a loss of efficiency and lead to increased *boiler scale* buildup inside the collectors.

Automatic draining to provide freeze protection is accomplished by using a differential thermostat with a separate control circuit to activate two 2-way solenoid valves. There is a *winter-summer* switch on the thermostat. When in the winter position, the collectors are *automatically* drained at the end of each day as the pump motor shuts off. This discharges approximately 3 gallons of water for a 3-collector system (about the amount used in flushing a toilet). With the switch in the *summer* position, the solenoid valves are electrically disconnected. A shut-off valve on the drain must be manually closed when you decide to switch to summer operation. This is necessary, because the solenoid drain valve is of the *normally open* type, which means that with no electrical impulse to the valve, it is *open* to the drain. The use of a normally open valve provides automatic draining in winter weather even if a power failure should occur. The summer mode of operation conserves water during the period when freezing weather is not present. An air-relief valve is

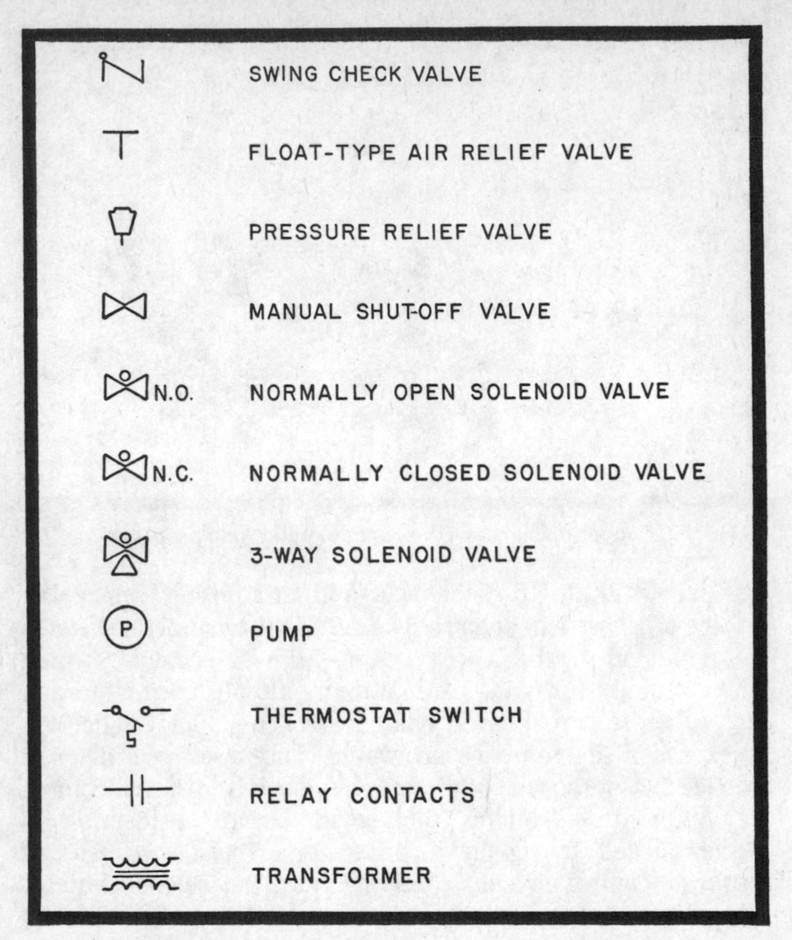

Fig. 7-2. Symbols used in schematic drawings of solar energy systems.

installed at the highest point in the plumbing line to facilitate draining and to allow air to escape when the collector loop is refilled with water.

Figures 7-2 and 7-3 will help you understand the overall system function. Symbols used in the schematic drawings of Chapters 7, 8, and 9 are shown in Fig. 7-2. Figure 7-3 provides a schematic for the system highlighted in this chapter.

In addition to controlling automatic draining and overly high temperatues, the Rho Sigma model RS 500 1PH-2L-AL differential thermostat starts and stops the Grundfos UP 25-42 SF circulator pump and varies its *speed* as a function of the temperature difference between the absorber and storage sensors. The pump will run slowly until the temperature difference increases, then the pump speed (and flow rate) will increase to improve the overall efficiency of the system. The theory of variable flow control is explained in more detail in my *Homeowner's Guide to Solar Heating & Cooling*. You may also wish to refer to Chapter 1 and Fig. 1-2 of this book for an explanation of the normal function the differential thermostat performs in starting and stopping the pump.

In this circulator pump and drain-down system, an 82-gallon *storage-only* tank is used. It has no heating element and serves as a preheater for your existing water heater. Separate storage allows hot water to be collected independently of its use; it eliminates the need to alter your normal daily routine of washing clothes, dishes, and bathing; and it provides improved system efficiency. I chose the model 666H-82-T manufactured by Rheem. If you select another unit, I would suggest it be designed for solar operation with an

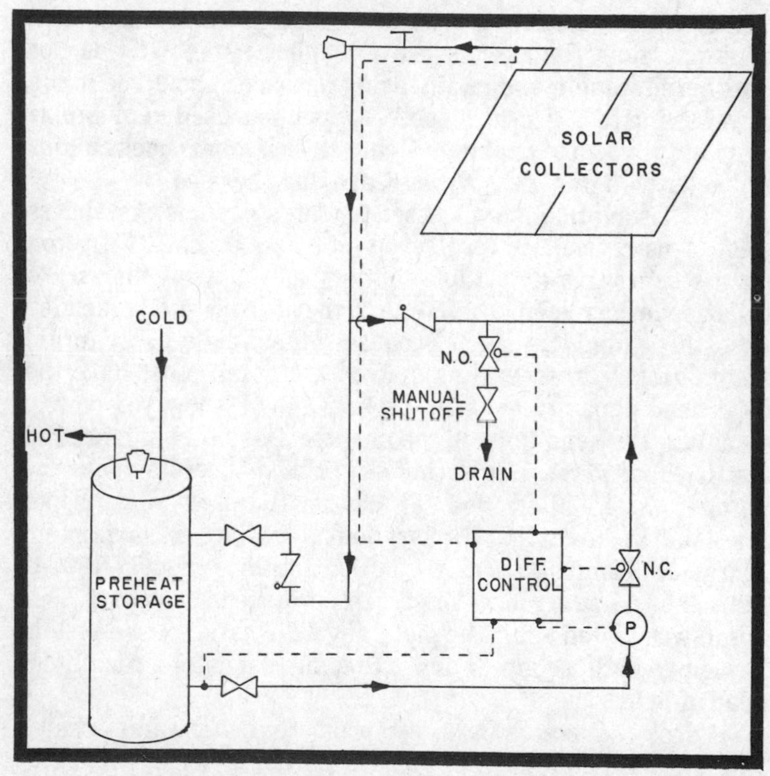

Fig. 7-3. Schematic representation of drain-down system.

elevated feed line to reduce scale and sediment buildup. It should also have a collector return line on the *side* of the tank located approximately one-third of the way down.

Collector construction is similar to the Chapter 6 design with two exceptions. The risers in this system run longitudinally or parallel to the long dimension of the collector. This means the collectors must be mounted with their short dimension positioned horizontally. This change eliminates much of the soldering on the absorber tubes, as 6 *long* risers are used rather than 12 *short* ones. In addition to less labor and fewer joints, I estimated the material savings to be $20 per collector. Also, the inlet and outlet connections to the collectors are on *opposite* corners to provide more uniform flow. When this design is connected using reverse return piping (look ahead to Fig. 7-6), equal distribution of water to each collector is theoretically the same, eliminating the need for balancing valves. In the thermosiphon system of Chapter 6, the idea was to keep the lines as short as possible and to minimize flow restrictions, so this piping method was not considered. Single and dual glazing decisions should be made by referring to Fig. 5-10. Type M copper was used to minimize cost; but if you are unable to locate it, or if local codes call for a heavier wall thickness, types K or L may be used.

The circulation pump chosen for this system is of stainless steel construction to protect it from corrosion. Cast-iron pumps circulating potable water have been known to experience corrosion problems and they are not recommended for this application. Because the collectors drain automatically in freezing weather, the pump must have the total head capacity to refill the lines. I will show you how to calculate the head later in the chapter. Basically, the head in feet is equal to the height (in feet) of the collectors above the pump *plus* an allowance for frictional losses in horizontal lines, and in valves and fittings, before reaching the maximum height. A performance curve on the Grundfos model UP 25-42 SF is shown in the manufacturer's literature in Chapter 3. This pump was tested using the plumbing schematic for the system described in this chapter and is recommended for a maximum head of 14 foot.

Three collectors are required to satisfy the design example that follows, but as in Chapters 6 and 8, the *bill of materials* will always be based on a 2-collector system so that you can make a direct comparison of what the different designs cost to make.

DESIGN EXAMPLE

Following are the calculations used to develop design parameters:

Assumed load:

hot water service temperature ....120° F
daily hot water usage for family of three with washing machine....................80 gallons
average daily outdoor temperature (annual) ....62° F
average monthly load............1,199, 626 Btus

Solar energy available (average month)

location ....Los Angeles
latitude........................34°
collector tilt and orientation ......34° facing south
collectible energy (Btus per month per square foot) 56,662

Collector design

number of collectors 3
number of covers....................1
cover material..3/16 inch thick tempered glass
absorber material................1/2 type M copper risers soldered to 10-ounce copper sheet
absorber coating ....flat-black paint
insulation ....industrial grade fiberglass
frame ....extruded aluminum
nominal size ....36″ × 78″ × 4″

Storage tank (Rheem 666H-82-T or equal)

capacity....................82 gallons
heating element ....none
weight ....248 pounds

MATERIALS: AVAILABILITY AND COST

The materials that follow cost approximately $910 as of this writing. Chapters 2 and 3 provided detailed information should you wish to make substitutions. No return or feed lines are included in the estimate, because the length and complexity of these lines will vary from job to job. Make your own estimate from local sources of supply, then decide on the economic advantages of proceeding with construction.

Bill of Materials for a Two-collector System

Quantity	Size	Description
Absorber:		
4 pcs.	1/2 in. diam. × 20 ft	type M copper water tube (0.625 O.D.), hard temper, straight lengths
1 pc.	1 in. diam. × 10 ft	type M copper water tube 1 in. (1.125 O.D.), hard temper, straight lengths
2 pcs.	0.021 × 32″ × 74″	10 oz (0.014 in. thick) copper sheet in cold rolled temper
20 pcs.	1 × 1 × 1/2	copper tees, C × C × C wrought solder-type pressure fittings
4 pcs.	1 × 1/2	copper 90° ells, C × C wrought solder-type pressure fittings
4 pcs.	3/4 × 3	copper pipe with 3/4 in. NPT pipe threads one end
4 pcs.	1 × 3/4	copper 90° ells, C × F cast solder-type pressure fittings
Collector frame:		
4 pcs.	10 ft	aluminum frame extrusion, Alcoa die number 459831
4 pcs.	10 ft	aluminum glazing cap extrusion, Alcoa die number 459841
1 pc.	2 ft	aluminum hold-down clip extrusion, Alcoa die number 459851
60 pcs.	No. 10 × 3/4 in.	No. 10 type A, round or pan head sheet-metal screws, 3/4 in. long, cadmium plated steel

Quantity	Size	Description
Back plate:		
2 pcs.	1/4″ × 4′ × 8′	exterior grade or marine plywood, 1/4 in. thickness
Collector insulation:		
14 pcs.	1″ × 24″ × 48″	industrial grade semi-rigid fiberglass board (Certain-Teed Corp. No. 850, Johns-Manville No. 814 or cqual), no facings
Glazing:		
2 pcs.	3/16″ × 34″ × 76″	tempered glass with low iron content
1 roll	1/4″ × 1/4″ × 50′	General Electric Silglaze tape SCT 1534 or equal
1 cartridge	1/12 gal	General Electric Silglaze silicone sealant cartridge SCS 1803 or equal
1 cartridge	1/12 gal	General Electric 1200 series silicone sealant cartridge SCS 1203 or equal
1 doz	1/8″ × 1/8″ × 1/2″	hard neoprene rubber setting blocks
Valves:		
2	3/4″	isolation valves—may be either gate or ball type. Cast brass with 3/4 in. NPT threads both ends
2	3/4″	flanged isolation valves for use with Grundfos pump. Part No. UP 25-42 SF
2	3/4″	swing check valves. Cast brass with 3/4 in. NPT threads both ends. Minimum 100 psi rating
2	3/4″ × 3/4″	pressure relief valves. 100 psi pressure setting, 3/4 in. NPT threads

Quantity	Size	Description
1	3/4″	automatic air vent valve with 3/4 in. male NPT threaded connector. 150 psi rating. Maid-O'-Mist No. 75 Auto-Vent or equal
1	1/4″ × 1/4″	2-way solenoid valve. Normally open 24 volt. 1/4 in. NPT threads. ASCO model 8262262 or equal
1	3/4″ × 3/4″	2-way solenoid valve. Normally closed 24 volt. 3/4 in. NPT threads. ASCO model 8210D95 or equal
1	1/4″ × 1/4″	brass shut-off valve (for manually closing collector drain during summer operation
Control system:		
1		differential thermostat with proportional flow control, high temperature turn-off and 2 24-volt outputs for solenoid valve control. Rho Sigma model RS 500 1PH-2L-AL or equal
2		temperature sensors. Rho Sigma model SP or equal (for measuring pipe fluid temperatures)
Pump:		
1	1/20th hp	circulator pump. Stainless steel construction with sufficient head capacity for your specific installation. Grundfos model UP 25-42 SF or equal

Quantity	Size	Description
Miscellaneous parts and plumbing:		
4	3/4″ × 3/4″	brass unions, C × C wrought solder-type pressure fittings (used to join collectors to plumbing lines)
2	3/4″I.D.	CPVC thermoplastic couplings. Unthreaded schedule 80. (To insulate absorber inlet and outlet fittings from collector frame. This part is cut into 4 equal pieces.)
*2	3/4″ × 3/4″ × 1/2″	brass tees, C × C × F cast solder-type pressure fittings (to connect temperature sensors into plumbing lines)
*1	3/4″ × 3/4″ × 3/4″	brass tee, C × C × F cast solder-type pressure fitting (to connect air vent valve to collector return line)
*1	3/4″ × 3/4″ × 3/4″	brass tee, C × F × C cast solder-type pressure fitting (to connect pressure-relief valve to collector return line)

**Note:* Only a few of the *special* fittings are itemized. These are based on 3/4-inch copper lines and solder-type connections. You may decide to use *threaded* connections, and this is satisfactory provided 3/4-inch collector feed and return lines are maintained. The balance of the fittings you require for your specific job may be selected using Chapter 2 as a guide.

Quantity	Size	Description
Storage tank:		
1	82 gal	Rheem model 666H-82-T or equal
General items:		
1 qt		heat-resistant flat-black paint
3 lbs	1/8 in. diam.	50A wire solder (50% tin, 50% lead by weight)

Quantity	Size	Description
1/2 lb		good quality paste flux. LA-CO (Lake Chemical Co.) or equal
as needed	3/4″ I.D.	closed-cell foamed rubber tubing insulation. 1″ minimum wall thickness (to insulate connecting lines)
as needed		adhesive to cement foamed rubber tubing insulation

ABSORBER FABRICATION AND ASSEMBLY

The *absorber* is the only basic difference between this collector and the one described in Chapter 6. The following description, therefore, will not repeat all the fabrication and assembly techniques, but will provide the dimensional changes and figures needed to fabricate *one* collector.

Cut two 20-foot lengths of 1/2-inch tube into a total of 6 equal pieces, 6 feet long. There will be four feet of scrap left over—two from each tube. This tubing *is* available in 20-foot lengths, but more outlets stock 10-foot pieces. If you are unable to locate the longer length, you would have 24 feet of scrap instead of 4 feet because you would need six lengths and each would yield four feet of scrap. An option would be to splice some tubes with 1/2″ × 1/2″ C × C couplings.

Next, cut the 1-inch tube into 10 pieces that are 5 inches long, and two pieces 2 inches long.

Soldering the Absorber Plate

Lay out the pieces you have cut along with the tees and ells, positioning them as shown in Fig. 7-4. Clean, flux, and solder as previously described. Trim the 10-ounce copper sheet to 33″ × 72″, using tin snips. Place the tube assembly on top of the sheet with the 72 inch sides just touching the headers and a 1 1/2 inch overhang on each side of the outside risers. Mark, clean, flux, and solder the parts together.

Pressure Testing and Painting

Refer to Chapter 6 for a description of these operations.

FRAME FABRICATION

Fabrication of the frame is identical to that for the Chapter 6 collector with one exception: one 1 1/4 inch diameter

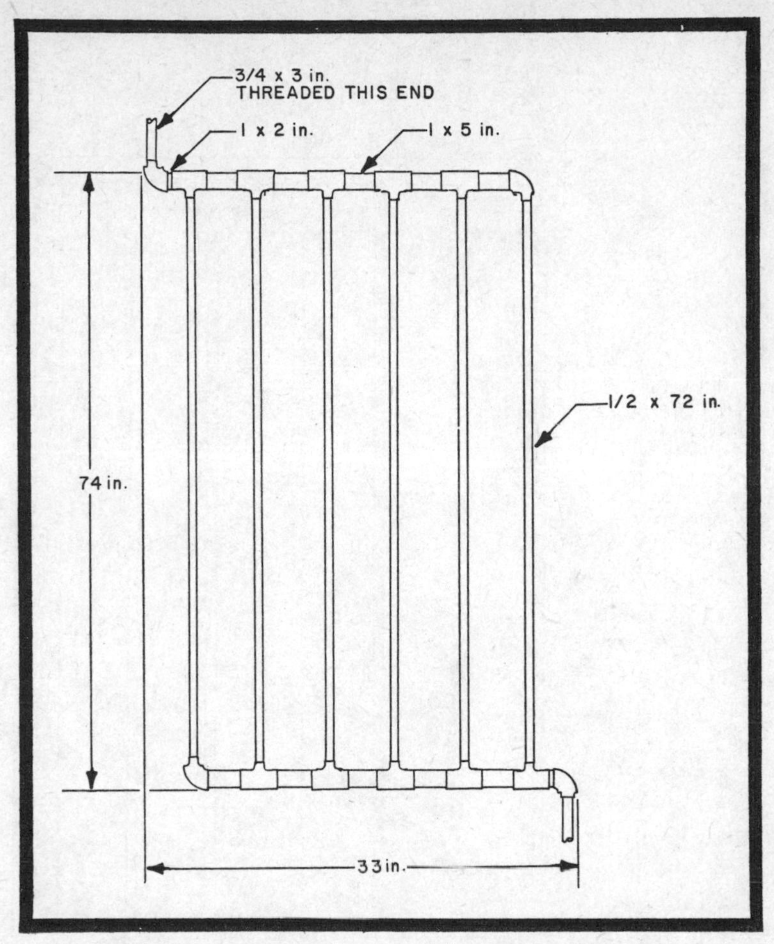

Fig. 7-4. Assembly diagram of absorber components.

hole for the inlet and outlet headers is made in each of the 35 3/8 inch long frame pieces.

FRAME AND BACK PLATE ASSEMBLY

Due to the fact that the inlet and outlet fittings are on opposite ends of the collector, it will be necessary to leave one 3/4 inch threaded pipe off the absorber until the unit is installed in the collector frame. Position the absorber (see Fig. 7-5), then install the 3/4 inch pipe using wet thread sealer or Teflon tape on all threaded connections. It is critical that there be no leakage at these joints, and I would recommend that you pressure test again before installing the glazing.

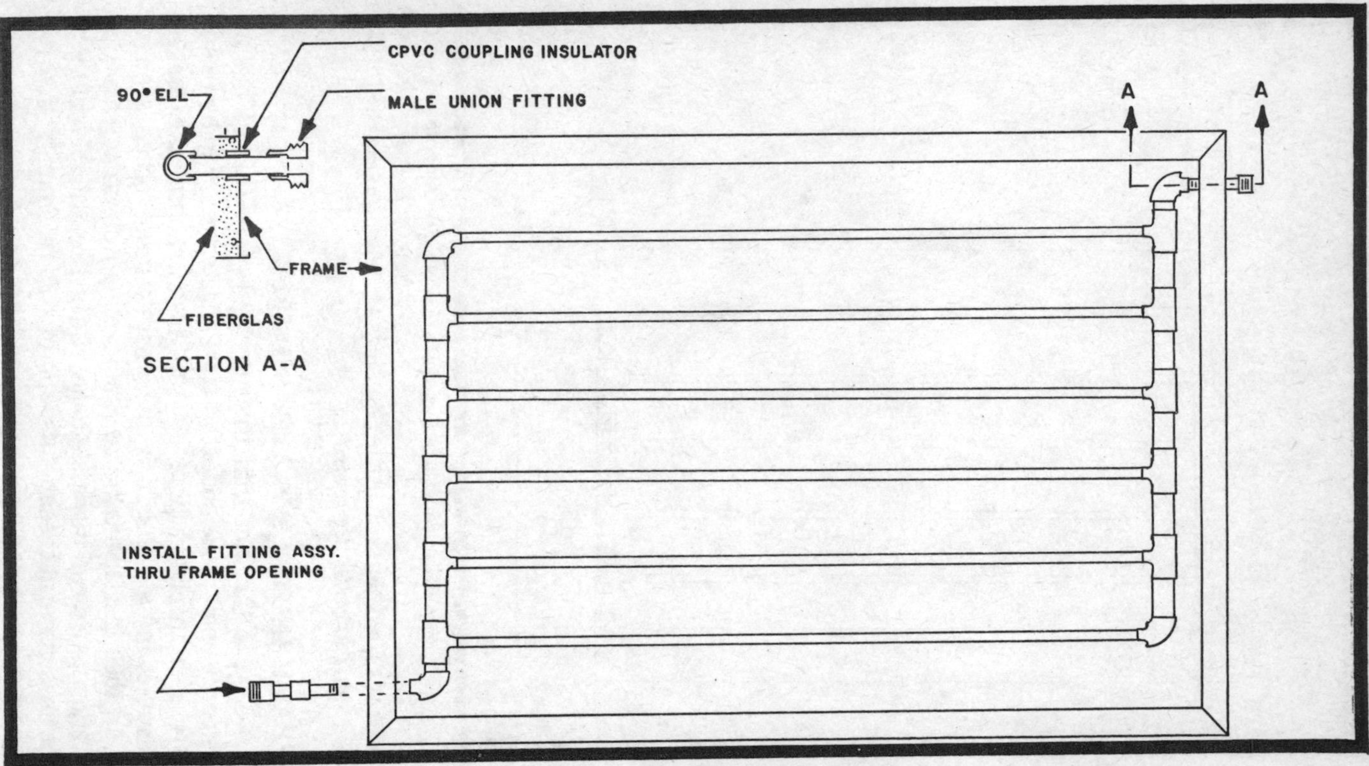

Fig. 7-5. Method of installing absorber in collector frame.

COLLECTOR LOCATION

The collectors should face approximately south, and should be tilted at the angle of latitude. A plus-or-minus 10° deviation from this angle will not affect the performance very much. It is most important to select a location that is unshaded during the prime collection hours of 9:00 a.m. to 4:00 p.m. over the entire year, and one where the length of lines connecting the collectors to the storage tank can be kept to a minimum. Use a method of mounting that satisfies your local code regarding structural loads, including wind, hail, and snow if applicable. Each collector should weigh approximately 100 pounds when filled with water.

PLUMBING CONNECTIONS

Figure 7-6 provides a detailed plumbing diagram of the system. The two C × C × F tee fittings at points A and B have a 1/2 inch threaded opening for temperature sensors. Sensors will be installed later, using the method suggested in the electrical section that follows.

Two swing-type check valves are used in the lines. The one at point C is mounted *vertically*. It prevents reverse flow through the system at all times and losses from the preheat storage tank when the solenoid drain valve is open. The check valve at point D is installed in the *horizontal* drain bypass line with its hinge or pivot on *top*. While the pump is operating, the *pressure* on the pump side of the valve is high enough to keep this valve *closed* and prevent the heated water leaving the collectors from mixing with the cooler water on the pump side.

The important points to remember about the combination vent-vacuum relief valve are that it must be installed in the *vertical* position and at the *highest* point in the plumbing line. This valve is float operated. When the valve shell is full of water, the valve is closed. When the circulation pump stops and the drain valve opens, the float drops, allowing air to enter the lines and facilitate draining. The valve is installed at point F by threading it into a 3/4″ × 3/4″ × 3/4″ C × C × F tee. The F, or female portion of the tee, should be on top of the line and vertical.

It is important to locate the differential thermostat, pump, and both solenoid valves *inside* the house, out of the weather. If there is no convenient waste line near the storage tank for draining the collectors, you may consider connecting into the

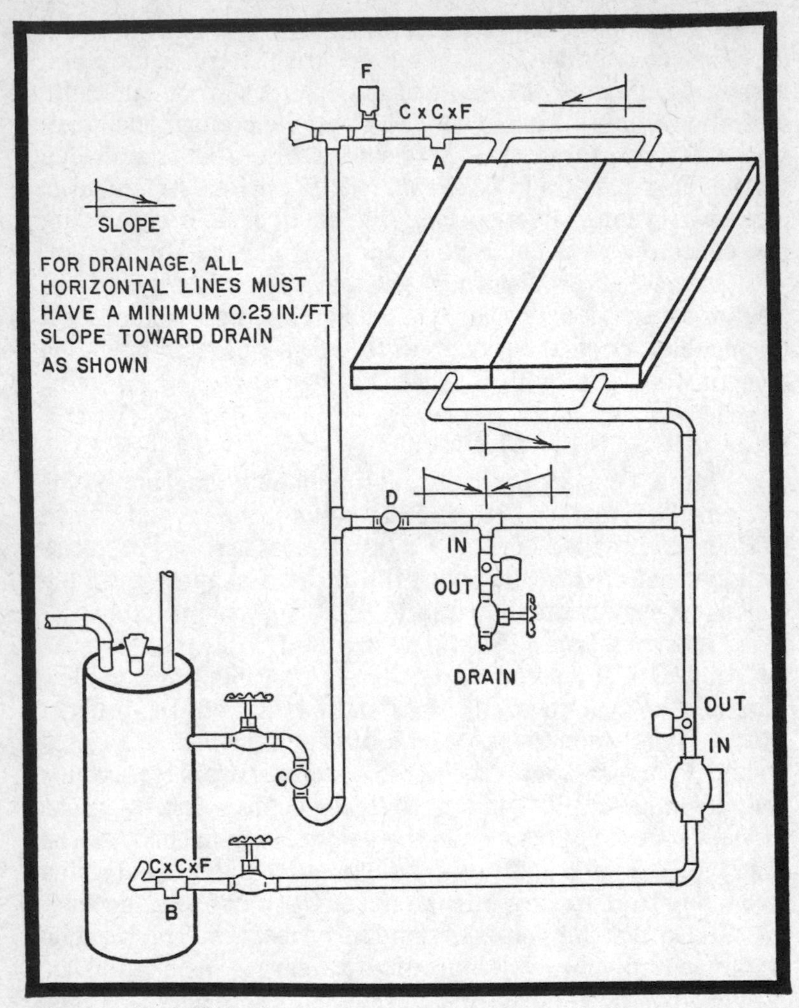

Fig. 7-6. Plumbing diagram of drain-down system. Note importance of sloping lines toward drain.

drain line leaving the pressure-relief valve on either the storage tank or water heater tank. The solenoid valves are installed in the approximate positions shown. Each valve is clearly marked "in" and "out" on the valve body. Be sure to install them in the right direction. If you use pipe compound when installing valves, apply it to male threads only, not to the valve threads. I prefer to use Teflon tape to seal pipe joints, as there is less chance that this material will get inside the valves and cause problems.

Both the pump and solenoid shut-off valve are installed in the collector feed line *above* the level at which this line enters the storage tank. This helps to protect both components from sediment which may be in the line. Isolation valves on each side of the pump are recommended by the manufacturer and are a good investment. It is also advisable to support the plumbing lines near the pump to prevent undue stress on the plumbing.

Be sure the drain bypass line is sloped and components are positioned according to the layout in Fig. 7-6. When assembling the parts, I suggest starting at the storage tank and working toward the collectors on both lines, as the collectors are attached with unions. Clean and flux the joints to be soldered; but before applying any heat, be sure that any electrical components such as the top of solenoid valves, sensors, etc. Are removed; heat may damage them. It is also a good idea to install the air vent and relief valve after soldering. When the parts are in alignment (air vent and swing check vertical, etc.), solder the parts together.

CALCULATION OF HEAD LOSS

Because the pump must refill the collectors each time they drain, it must have adequate capacity, referred to as *feet of head*. Head losses are the total loads placed on the pump by resistance to fluid flow through valves, fittings, solar collectors, and horizontal lines, and the vertical distance in feet to which the pump must raise the water. If the losses are less than the pump's capacity, the collectors will refill. Once the lines are full, the pump has a much easier job, because water traveling back to storage creates a siphoning effect and the pump must merely circulate it.

The desired flow rate for this system is 1 gallon/square foot of collector per hour. For a 3-collector system, the effective absorber area is 32″ × 74″ per collector—or a total of 49 square feet yielding a flow rate of 0.82 gallon/minute. At this low rate, you will find the major loss will be vertical height. In fact, published data from most suppliers do not provide values for pressure loss due to friction (head loss) at flow rates below 1 gallon/minute, so you should base calculations on the 1 gallon/minute rate which is conservative.

Table 7-1 lists frictional losses for common sizes of copper tube, per 100 feet, and Table 7-2 gives fitting values. To give

Table 7-1. Friction Loss per 100 ft of Type L Copper Tube Due to Water Flow.

Nominal Tube Size	Pressure Drop (psi)	Lost Head (ft)	Flow Rate (gpm)
1/2″	1.1	2.5	1
	3.8	8.8	2
	7.7	18	3
	13	30	4
	19	44	5
	26	60	6
3/4″	0.20	0.46	1
	0.69	1.6	2
	1.4	3.2	3
	2.2	5.1	4
	3.4	7.8	5
	4.6	11	6
1″	0.06	0.13	1
	0.19	0.44	2
	0.40	0.92	3
	0.62	1.4	4
	0.93	2.2	5
	1.3	3.0	6

you an example of the head calculation: assume your system uses 40 feet of 3/4 inch copper connecting lines, 4 tee runs (straight through), two 90° ells, and 2 tees with side outlets. Refer to Tables 7-1 and 7-2.

40 feet of 3/4″ tube ....0.18 feet
4 tee runs = 2 feet of tube
2 90° ells = 2 feet of tube
2 tee side outlets = 4 feet of tube
total fittings = 8 feet of tube = 0.04 feet
total loss due to connecting tube and fittings ..0.22 feet

At the recommended flow rate, the pressure drop across each collector is quite small—in the neighborhood of 0.1 foot. From

Table 7-2. Approximate Friction Loss Allowances for Fittings and Valves in Feet of Straight Tube.

Nominal Size of Tube (Inches)	90° Elbow	45° Elbow	Tee, Run	Tee, Side Outlet	90° Bend	180° Bend	Compression Stop
1/2	1/2	1/2	1/2	1	1/2	1	
3/4	1	1/2	1/2	2	1	2	
1	1	1	1/2	3	1	2	
Cast Brass Fittings and Valves							
1/2	1	1/2	1/2	2			13
3/4	2	1	1/2	3			21
1	4	2	1/2	4			30

these calculations, it should be apparent that for these low flow rates the controlling factor in pump selection is the vertical rise. As the Grundfos model UP-25-42 SF pump selected for this system can handle 12 feet of head at this flow rate, you should have no problem with most installations. If your specific job has a higher head, a larger pump or two pumps in series may be required.

ELECTRICAL CONNECTIONS

The model RS 500 1PH-2L-AL differential thermostat manufactured by Rho Sigma has been specially engineered for this application. The "1PH" denotes that the unit has one proportional motor control (P) and one high temperature turn-off circuit (H). The 2L designation indicates two *low level* outputs (L) for 24-volt solenoid valve operation. In addition to the *on-off-auto* switch, a *winter-summer* operating switch is included in the control box.

In connecting the thermostat, there is a terminal strip inside the control box with clear identification of outputs and inputs. You should not have difficulty with this step, but be sure to use the proper wire size and type in connecting the

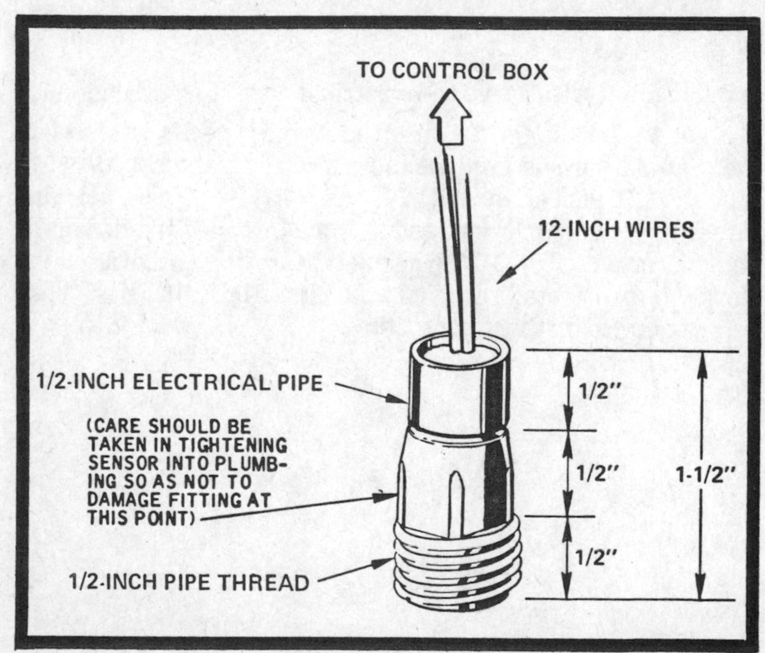

Fig. 7-7. Sensor used to measure fluid temperature in pipes (SP type).

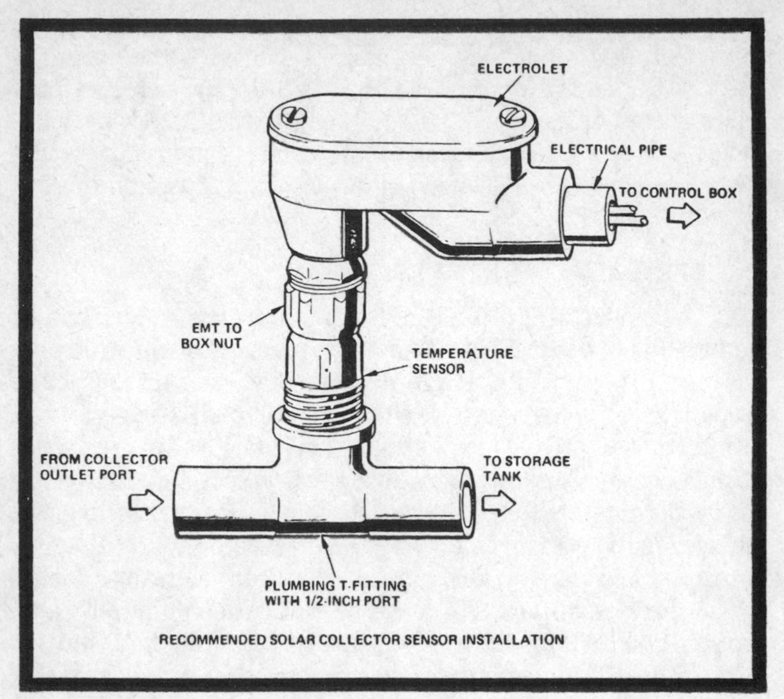

Fig. 7-8. Method of installing an SP type sensor suggested by Rho Sigma, Inc.

components and observe the electrical codes and precautions.

The sensors suggested are Rho Sigma type SP, as shown in the manufacturer's product literature in Chapter 3. Observe that the SP sensor in Fig. 7-7 has a 1/2 inch national pipe thread. The units are screwed into the threaded tee fittings at points A and B (Fig. 7-6). When necessary to weatherproof the sensor, the installation shown in Fig. 7-8 has been recommended by the manufacturer.

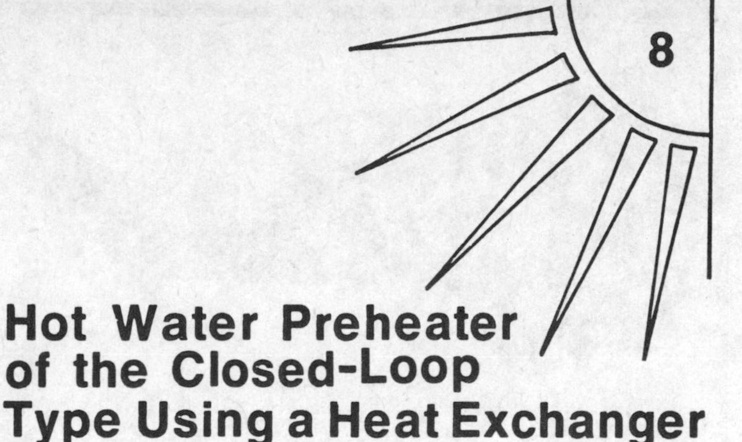

Hot Water Preheater of the Closed-Loop Type Using a Heat Exchanger

With this closed-loop system, there are no geographic restrictions because freeze protection is provided—only sunshine is required. You will appreciate the simplicity of construction of this system. But because the unit is more expensive than the systems described in Chapter 6 and 7, it is specifically recommended for use in only the "D" region of Fig. 5-2, or that portion shaded in Fig. 8-1, where freezing weather occurs. Dual glazing is recommended here, as was shown in Fig. 5-10.

SYSTEM DESCRIPTION

A compact, prepackaged pumping, heat transfer, and control module made by Taco, Inc. is used in this design. This unit includes two pumps, a heat exchanger, differential thermostat, sensors, expansion tank, and shut-off valves. The module is factory-assembled, requiring only one electrical and four plumbing connections; it is pictured in the manufacturer's literature printed in Chapter 3. The name *closed loop* system is used because the fluid circulating through the collectors never comes into contact with the potable water in the storage tank. The liquids are separated by a *heat exchanger* in the control module. Each liquid is circulated by a separate pump. The differential thermostat which controls both pumps automatically starts circulation when the collector temperature exceeds storage by 20° F. When the temperature difference drops to 3° F, the pumps turn off. Figure 8-2 shows the system diagram. All parts within the heavy dotted line labeled SSM are included in the control module.

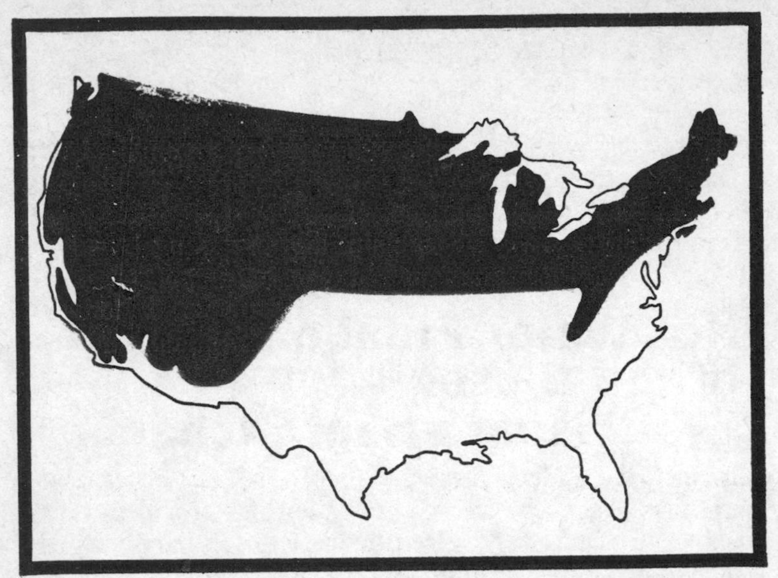

Fig. 8-1. Suggested climatic zone for closed-loop water heating system.

In this design, the collectors are not subjected to water supply line pressure, so there should be no problem in using type M copper tube to build and connect them. The *storage loop*, however, operates at line pressure, and local plumbing codes must be observed. Relief valves are installed in both loops. They are designed to open automatically, to discharge, and then to close once excess pressure has been relieved. An air vent valve is installed at the highest point in the system. When the valve shell is full of fluid, the valve is closed. If air accumulates, the float drops, opening the valve and allowing the air to escape. Water again fills the shell, raising the float and closing the valve. The collectors and exposed plumbing lines are protected against freezing weather by circulating a buffered 50/50 ethylene glycol (antifreeze) and water mixture, with a freezing point of approximately −34° F, in the collector loop.

Separate storage is designed into this system. By connecting the storage tank in series with a conventional water heater, hot water may be collected independently of its use. In addition to this feature, separate storage in a well-insulated tank provides greater efficiency by allowing the collectors to operate at a lower temperature. For example, the average temperature in the *storage* tank may be 100° F and this is *solar* energy. In the conventional water heater, the average temperature is slightly above the thermostatic temperature you select, say 125° F. The difference of 25° F is provided by *gas* or *electricity*. The solar systems operating in this manner are properly called *preheaters*.

This is the first time I have introduced the term *heat exchanger*, and it requires some explanation. The types most frequently used in solar systems are the following: air-to-liquid, liquid-to-air, and liquid-to-liquid. These designations describe the medium through which the heat is transferred. A *solar collector* is an air-to-liquid exchanger, because it transfers solar energy to the fluid flowing through its absorber. The exchanger located in the furnace *plenum* (Chapter 9) is of the liquid-to-air type. For the closed-loop

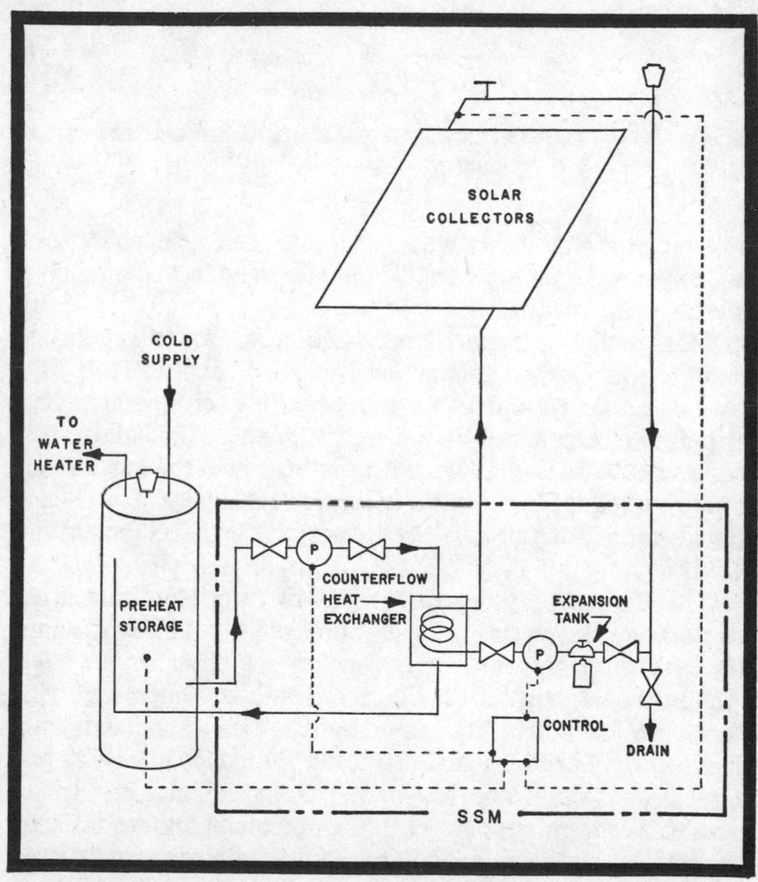

Fig. 8-2. Schematic diagram of closed-loop water heating system.

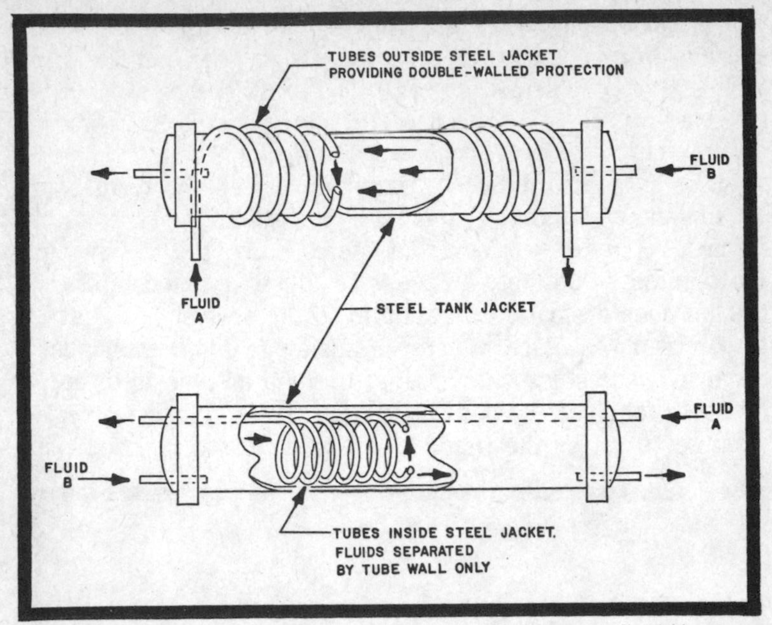

Fig. 8-3. Comparison of single- and double-walled liquid-to-liquid heat exchangers.

system in this chapter, however, a *liquid-to-liquid* exchanger is used to separate the fluid in the collector loop from the potable water in the storage tank.

Most building departments and local codes require a *double-walled* heat exchanger for this application. The reasoning is that if a *leak* were to occur inside the exchanger, the potable water would mix with the collector fluid and might pose a health hazard. Figure 8-3 pictures both the single- and double-walled types of liquid-to-liquid exchangers. Taco has a double-walled, or *fail-safe*, exchanger under development for its SSM

This design provides more latitude in collector selection than other designs do. Because the fluid circulating through this system is not potable water, copper absorbers are not mandatory as long as the fluid selected is compatible with aluminum or steel. The manufacturer's recommendations should be followed. I chose to build the collectors using the same design described in Chapter 7 and keeping the entire plumbing loop in copper. With copper plumbing required on the potable water side and the heat exchanger constructed using copper tube, I am of the opinion that the extra expense of using all copper is justified to minimize the possibility of galvanic corrosion due to dissimilar metals. If you elect to purchase a commercial collector rather than building your own, I would recommend that the fluid passages be copper. Dual glazing is suggested in accordance with information in Fig. 5-10. The collectors are connected as in Chapter 7, using reverse return piping. This insures equal distribution of the pumped fluid to each collector without the need for balancing valves.

Using a heat exchanger and fluid other than water in the collectors lowers the efficiency of the system, but dual glazing on the collectors helps keep the losses down. If you are on the borderline between using 4 and 5 collectors (based on your individual load and climate), go for the extra collector.

DESIGN EXAMPLE

Following are the calculations used to develop design parameters:

Assumed load:

hot water service temperature ....120° F
daily hot water usage for family of four (no dishwasher or washing machine) ....80 gallons
average daily outdoor temperature (annual) ....51° F
average monthly load ....1,427,141 Btus

Solar energy available (average month)

location ....Boston
latitude........................42°
collector tilt and orientation ......52° facing south
collectible energy (Btus per month per square foot) 41,318

Collector design

number of collectors 5
number of covers....................2
cover material ......insulating glass unit. 2 pcs. 1/8″ thick tempered glass with 1/2″ air space
absorber material...1/2 inch type M copper risers soldered to 10-ounce copper sheet
absorber coating ....flat-black paint
insulation ....industrial grade fiberglass
frame ....extruded aluminum
nominal size ....36″ × 78″ × 4″

Storage tank (Rheem 666H-120-T or equal)
capacity ....120 gallons
heating element ....none
weight ....399 pounds

Heat transfer and control module
Taco SSM........heat transfer and control module

Heat transfer fluid
collector loop ....50/50 solution (by weight) buffered eythylene glycol and water or 60/40 glycerol and water
storage loop ....potable water

MATERIALS: AVAILABILITY AND COST

The materials that follow cost approximately $1430 as of this writing. (Chapters 2 and 3 provide detailed information should you with to make substitutions). No return of feed lines are included in the estimate because the length and complexity of these lines will vary from job to job. Make your own estimate from local sources of supply, then decide on the economic advantages of proceeding with construciton. Note again that the following list is based upon a *two*-collector system, so that you will be able to make cost comparisons with other systems.

Bill of Materials for a Two-Collector System

Quantity	Size	Description
Absorber:		
4 pcs.	1/2 in. diam. × 20 ft	type M copper water tube (0.625 O.D.), hard temper, straight lengths
1 pc.	1 in. diam. × 10 ft	type M copper water tube 1 in. (1.125 O.D.), hard temper, straight length
2 pcs.	0.021 × 32″ × 74″	10 oz (0.014 in. thick) copper sheet in cold rolled temper
20 pcs.	1 × 1 × 1/2	copper tees, C × C × C wrought solder-type pressure fittings

Quantity	Size	Description
4 pcs.	1 × 1/2	copper 90° ells, C × C wrought solder-type pressure fittings
4 pcs.	1 × 3/4	copper 90° ells, C × F wrought solder-type pressure fittings.
4 pcs.	3/4 × 3	copper pipe with 3/4 in. NPT pipe threads one end
Collector frame:		
4 pcs.	10 ft	aluminum frame extrusion, Alcoa die number 459831
4 pcs.	10 ft	aluminum glazing cap extrusion, Alcoa die number 459841
1 pc.	2 ft	aluminum hold-down clip extrusion, Alcoa die number 459851
60 pcs.	No. 10 × 3/4 in.	No. 10 type A, round or pan head sheet-metal screws, 3/4 in. long, cadmium plated steel
Back plate:		
2 pcs.	1/4″ × 4′ × 8′	exterior grade or marine plywood, 1/4 in. thickness
Collector insulation:		
14 pcs.	1″ × 24″ × 8′	industrial grade semi-rigid fiberglass board (Certain Teed Corp. No. 850, Johns-Manville No. 814 or equal), no facings
Glazing:		
2 pcs.	3/16″ × 34″ × 76″	tempered glass with low iron content
1 roll	1/4″ × 1/4″ × 50′	General Electric Silglaze tape SCT 1534 or equal

Quantity	Size	Description
1 cartridge	1/12 roll	General Electric Silglaze silicone sealant cartridge SCS 1803 or equal
1 cartridge	1/12 gal	General Electric 1200 series silicone sealant cartridge SCS 1203 or equal
1 doz	1/8″ × 1/8″ × 1/2″	hard neoprene rubber setting blocks
Heat transfer and control module:		
1		Taco SSM heat transfer and control module
Valves:		
1	3/4″ × 3/4″	pressure-relief valve, 75 psi rating, 3/4 in. NPT threads
1	3/4″ × 3/4″	pressure-relief valve, 100 psi rating, 3/4 in. NPT threads
1	1/4″ male connection	air vent valve. Taco model 426 or equal
Miscellaneous parts and plumbing:		
4	3/4″ × 3/4″	brass unions, C × C wrought solder-type pressure fittings (used to join collectors to plumbing lines)
1	3/4″ I.D.	CPVC thermoplastic coupling. Unthreaded schedule 80. (To insulate absorber inlet and outlet fittings from collector frame. This part is cut into 4 equal pieces.)
*1	3/4″ × 3/4″ × 1/2″	brass tee, C × C × F cast solder-type pressure fitting (to connect temperature sensor to collector return line)
*1	3/4″ × 1/4″	brass 90° ell, C × F cast solder-type pressure fitting

Quantity	Size	Description
		(to connect air vent valve to collector return line)
*1	3/4″ × 3/4″ × 3/4″	brass tee, C × F × C cast solder-type pressure fitting (to connect pressure relief valve to collector return line)

**Note:* Only a few of the *special* fittings are itemized. These are based on 3/4″ copper lines and solder-type connections. You may decide to use *threaded* connections, and this is satisfactory provided 3/4 inch collector feed and return lines are maintained. The balance of the fittings you require for your specific job may be selected using Chapter 2 as a guide.

Storage tank:		
1	120 gal	Rheem model 666H-120-T or equal
General items:		
1 qt		heat-resistant flat-black paint
3 lbs	1/8 in. diam.	50 A wire solder (50% tin, 50% lead by weight)
1/2 lb		good quality paste flux. LA-CO (Lake Chemical Co.) or equal
as needed	3/4″ I.D.	closed-cell foamed rubber tubing insulation. 1″ minimum wall thickness (to insulate connecting lines)
as needed		adhesive to cement foamed rubber tubing insulation

COLLECTOR FABRICATION AND ASSEMBLY

The collectors used are identical to the ones described in Chapter 7. Construction and pressure testing should follow that outline.

ESTIMATING SYSTEM PERFORMANCE

In addition to the intensity of sunshine and outdoor temperature for your area, the temperature at which the

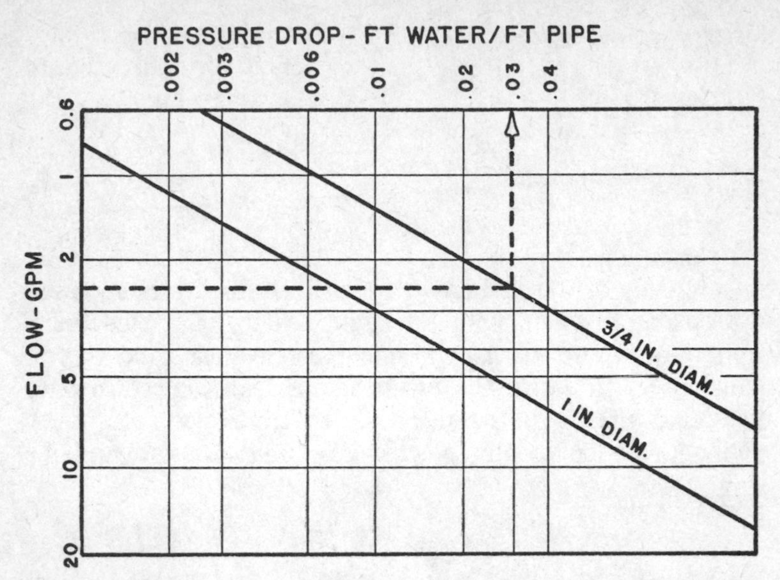

Fig. 8-4. Effect of frictional losses when circulating a 50 percent ethylene glycol solution.

storage tank is designed to operate has a dramatic effect on system efficiency. If you recall from Fig. 4-1, efficiency decreases as the temperature difference between the absorber and ambient temperature *increases*. Average absorber temperatures will vary almost directly with storage tank temperature as a function of the flow rate.

Instead of trying to store water at the *service* temperature of 120° F, consider designing for an *average* storage tank temperature of 100° F for this system. This does not mean the collectors will shut off when 100° F is reached. Quite the contrary, they will continue to collect energy as long as it is available. And during the summer, the tank should be able to provide service temperatures of 120° F without supplementary heat. When I am suggesting is that the tank *volume* be increased (120 gallons for this example) to keep the average tank temperature lower and the efficiency higher. The following calculations will help you to understand this concept.

For a 50/50 glycol-and-water heat transfer fluid, I recommend a flow rate of 0.5 gallons per minute (gpm) through each collector in the system. If you consider the 5-collector system suggested for the family of four living in Boston, the *total* flow through the system is 2.5 gpm (0.5 × 5). Now refer to the system diagram in Fig. 8-2 and calculate the *length of pipe* required to connect the collectors to the SSM module for your specific system. Be sure to include *both* feed and return lines. Add 30 percent to allow for valves and fittings. For this example, assume 50 feet of 3/4 inch copper tube was used. Adding 30 percent yields 65 *equivalent feet* of tube. There will be a pressure drop or *head loss* through the system as a function of the flow rate chosen, the diameter of plumbing lines, and the properties of the heat transfer fluid selected. Referring to Fig. 8-4, for a flow rate of 2.5 gpm and 3/4 inch I.D. copper tube, the pressure drop equals 0.03 feet of water for each *equivalent foot* of pipe in the system. Multiplying this figure by 65 yields a pressure drop, or *head*, of 1.95 feet.

Refer to Fig. 8-5 to determine the *operating point* of the system. This graph plots the performance curves of the Taco model 007 pump against the system curves using a 50/50 glycol-and-water heat transfer fluid. Continuing with our example, the flow is 2.5 gpm and the head is 1.95 feet. The values intersect one of the system curves at the *design point*

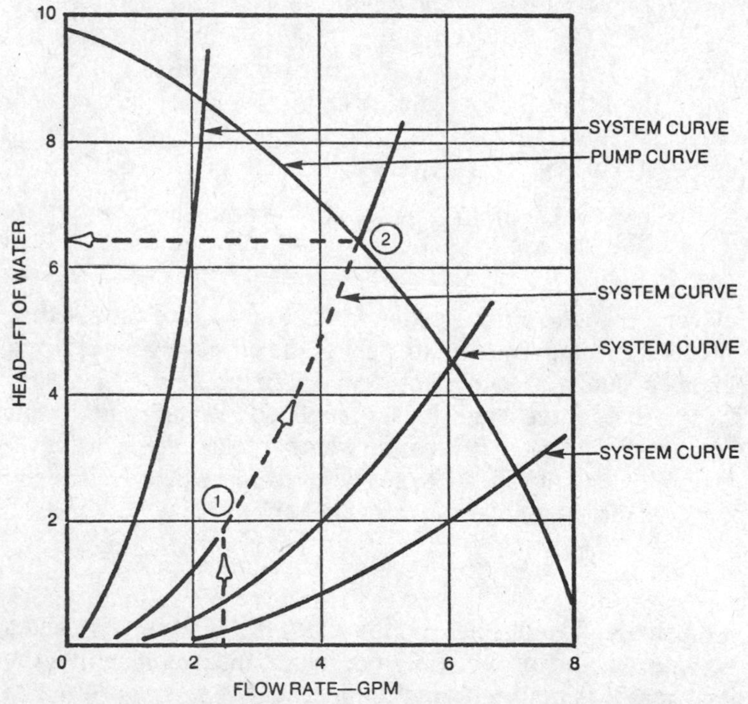

Fig. 8-5. Determination of pump capacity and operating point.

(point 1). Following the specific system curve to the point of intersection with the pump curve determines the actual point of operation (point 2). For this case, the flow will be 4.5 gpm at 6.4 feet of head.

For the parameters in the design example:

collectible energy = 41,318 Btus, per month per square foot
= 1333 Btus per day per square foot

Ninety percent of this energy is available in the middle two-thirds of the day, or during the 8-hour period from 8:00 a.m. to 4:00 p.m. The average *hourly* radiation, therefore, is:

$$\frac{0.90 \times 1333}{8} = 150 \text{ Btus per square foot}$$

For a 5-collector system having an area of 16.4 square feet of heat-collecting surface per collector and absorbing 50 percent of the available energy:

$$82 \text{ (sq ft)} \times 150 \text{ (Btu/sq ft)} = 0.50 = 6150 \text{ Btus}$$

You can now calculate the mean collector temperature by referring to Fig. 8-6. At 4.5 gpm flow rate and 6150 Btus per hour collected, the average absorber temperature will be 6° F higher than the storage tank. For the average daily outdoor temperature of 51° F in the design example (ambient temperature), the temperature difference, or ΔT, equals:

$$\begin{aligned} \Delta T &= \text{tank temperature } + 6° \text{ F} - \text{ambient} \\ &= 100 + 6 - 51 \\ &= 55° \text{ F} \end{aligned}$$

If you refer to the efficiency curves in Fig. 4-1, you can see that for this *low* temperature difference, the efficiency for either a single-or dual-glazed collector is approximately 60 percent. These are *average* conditions, though, so I would suggest that you check Fig. 5-10 to determine whether to use dual glazing.

Look again at Fig. 4-1 to see what happens to the efficiency if you decide to operate your storage tank at 120° F:

$$\begin{aligned} \Delta T &= 120 + 6 - 51 \\ &= 75°\text{F} \end{aligned}$$

For a single-glazed collector, the efficiency is 46 percent, and a dual-glazed unit provides 55 percent. This should prove the point that it is more efficient to operate the solar system as a preheater, which this design suggests.

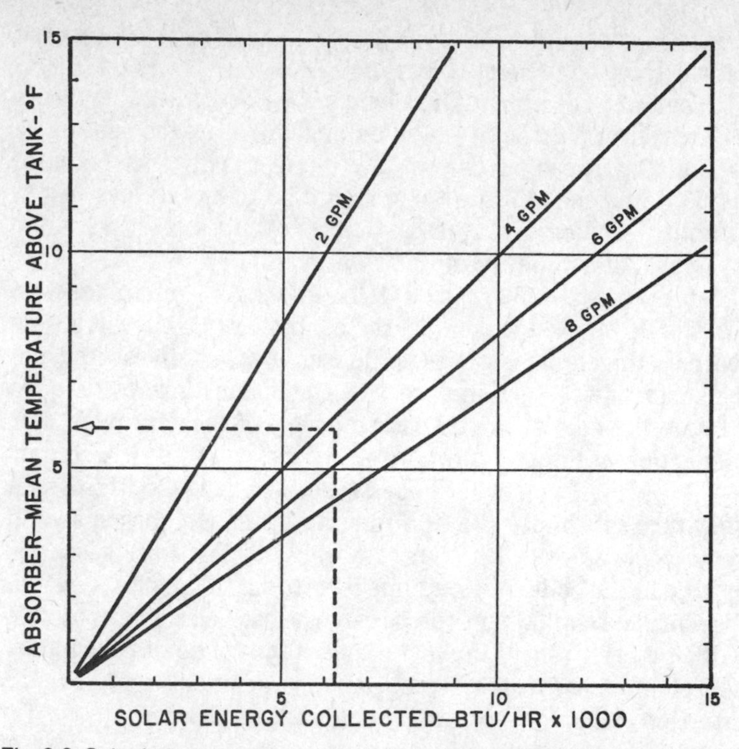

Fig. 8-6. Calculation of mean collector temperature for conditions chosen.

The assumption of 50 percent collector efficiency made earlier is too low for the operating conditions chosen. Adjusting to 60 percent:

$$82\ (\text{sq ft}) = 150\ (\text{Btu/sq ft}) \times 0.60 = 7380\ \text{Btus}$$

This means an *average* of 7380 Btus should be collected each hour the system is operating. In an 8-hour solar day, 59,040 Btus *may* be collected. It takes 1 Btu to raise one pound of water 1° F, so a 120-gallon tank (8.34 lb/gal) would have an average temperature increase of 59° F is no water was drawn from the tank and the plumbing lines and tank were *perfectly* insulated. In practice this is not possible, and when these factors are taken into consideration, the output of the 5-collector system should match the load in the design example with good efficiency.

INSTALLATION

The comments included in Chapter 7 on selecting a collector location would apply to this system as well. Plumbing connections should also be easy to understand, using reverse return piping to connect the collectors as shown in Fig. 8-2.

The Taco SSM module is designed for wall mounting by six 1/4-inch screws through the side flanges. Locate the unit inside the building in a dry location. If possible, the distance between the SSM and the storage tank should be 10 feet or less. When mounting, remember that the heat exchanger must be at the top to provide proper pump lubrication.

In wiring the unit, type THHN or #14AWG minimum wire size should be used to connect the controller to a 115-volt power source. Connection is made to the center pole of the switch and the open power terminal in the controller. Insert the wire through the hole at the transformer end of the controller. For connection of the two temperature sensors, use a twisted pair of #18 wires. If you order type *SP* sensors (see the Rho Sigma literature in Chapter 3), the mounting method chosen should be in accordance with Figs. 7-6 and 7-8. The *tank* sensor is located in the collector feed line next to the outlet on the side of the storage tank. Mount the *panel* sensor as closely as possible to the *outlet* fitting of the last panel in the collector return line. Sensor models *ST*, for direct panel attachment, and *SPI*, for insertion *inside* the storage tank, are also available.

INITIAL SYSTEM START-UP

The panel system may be filled through the hose drain connection below the entrance to the air scoop on the right-hand side of the panel. Simply open all valves on the panel portion of the module, attach a hose from the base connection to a suitable hand pump or jet pump, and pump from a sump up through the module and panel piping until the panel side is filled to the top and all air is forced out of the air vent. (Air absorbed under pressure will be removed by the air scoop and air vent during the natural sequence of circulation.) Continue pumping up the system until a pressure of 30 psi is attained. This will partially deflect the diaphragm in the sealed expansion tank, thus allowing for both expansion and contraction of fluid volume due to temperature variations. (The expansion tank is initially charged to 25 psi, but this may be increased up to 50 psi if the system requires it).

Under a 30 psi system pressure, the boiling point of a 50/50 mixture of ethylene glycol and water is about 295° F. However, if the system were to reach this temperature, the resultant expansion of the fluid would have increased the pressure, and thus the boiling point, so that boiling will not occur within the expected limits of operation of this system.

Adjustments of the flow should not be necessary on the panel side of the module if the selection procedure has been followed correctly. Variations in the flow velocity simply mean a lesser or greater gain in temperature through the panels as the flow is greater or less than design specifications may state.

MAINTENANCE AND SAFETY PRECAUTIONS

If you use ethlene glycol, buy a top grade that includes buffers or inhibitors to help prevent corrosion. This should be mixed in a 50/50 solution with distilled water and used as the heat-transfer fluid in the collector loop. Commercial antifreeze mixtures decay over a period of time, producing acidic concentrations. These acids are corrosive to copper and aluminum, and therefore the antifreeze solution must be drained and replaced periodically just as is required in an automobile radiator.

There is another option to ethylene glycol that I would *strongly suggest* you consider. A 60/40 percent solution of glycerol and water is nontoxic, noncorrosive, and has approximately 80 percent of the heat-transfer capabilities of a 50/50 ethylene glycol solution. Until a fail-safe exchanger is available, I would recommend that you plan for the lower efficiency and use thc glycerol as a safety precaution.

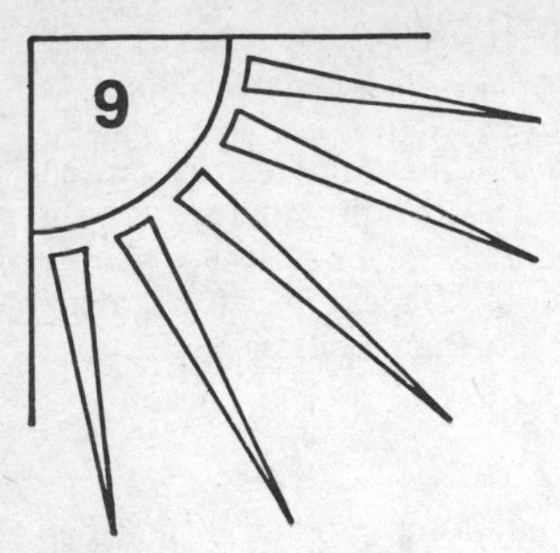

Solar Room Heater to Supplement a Forced - Air Heating System

This system offers some definite savings over professionally installed commercial units. It is similar to one used for demonstration purposes by the NASA (National Aeronautics and Space Administration) Langley Research Center in Hampton, Virginia. NASA does not endorse specific products, but you can obtain a copy of its system performance analysis of this unit on a 1500 square foot home by purchasing a copy of report number N76-27671, titled *An Inexpensive Economical Solar Heating System for Homes*, from the National Technical Information Service (NTIS) in Springfield, Virginia 22161. If, after reading this chapter, you plan to build a room heater of this type, I would urge you to purchase a copy of the NASA report. The price was $4.25 per copy as of this writing.

Initial costs for this system can be reduced by using polyvinyl fluoride (PVF) films for glazing. Other cost-cutting options include aluminum absorbers, CPVC high-temperature plastic pipe to connect the collectors, and an above-ground septic tank for storage. You should make your own decision on these options depending upon your budget, local climatic conditions, desired unit appearance, and the maintenance schedule you are willing to accept. From a durability standpoint, I suggest using the aluminum framing system for collectors and elevating them above the roof membrane a sufficient distance to provide drainage and reduce the growth of fungus, mold, and mildew.

The typical system described in this chapter is suggested for climatic zones 2, 3, 5, and 6 shown in Figs. 5-3 and 9-1. Dual glazing is recommended, as shown in Fig. 5-10. The home being considered should be well insulated (to FHA minimum standards), and should have an adequate unshaded area for roof-mounting the collectors. Also, it should have a suitable storage tank location. Installing the tank above-ground, outside the building, should only be considered in areas where single glazing is recommended.

SYSTEM DESCRIPTION

The system is designed to supplement an existing forced-air furnace. It will not carry the entire space-heating load throughout the year but can be economically sized to meet 40 to 60 percent of home heating requirements.

One circulating pump is used for the *dual* function of pumping inhibited water through the collectors and through a water-to-air heat exchanger located in the cold air return *plenum* of the forced-air furnace. Water flow is directed by two 3-way zone valves which are of the normally-open type, allowing the water to drain from the solar collectors by gravity whenever the collectors cannot contribute heat to the storage tank, or when there is a power failure. This feature provides freeze protection without the need for circulating a less efficient antifreeze solution. Draining is facilitated by a float-type air vent valve installed at the highest point in the collector return lines.

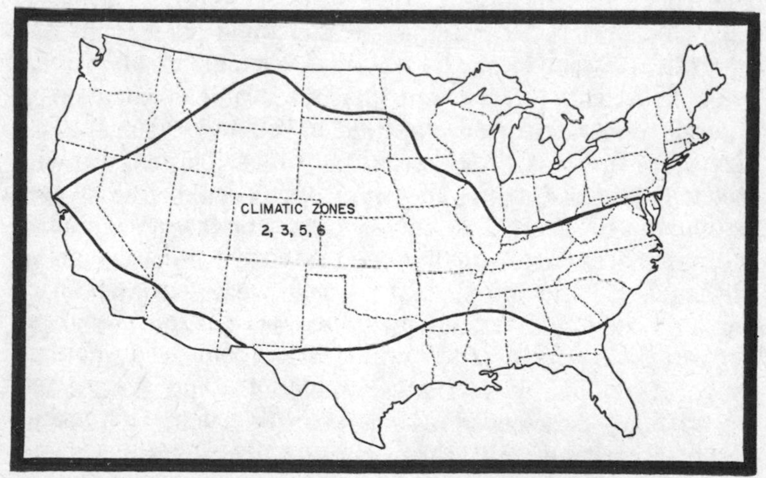

Fig. 9-1. Suggested climatic zones for home heating system.

The pump and zone valves are controlled by a differential thermostat, the house thermostat, and a thermostatic switch. On sunny days when the collectors are 10° F or more above the temperature of the storage tank, water flow will be directed through the collectors, heat exchanger, and into the storage tank. If the house thermostat calls for heat during this cycle, and the storage tank temperature is 75° F or above, the *blower fan* in the forced-air furnace will turn on, picking up heat from the exchanger and distributing it through the existing duct network. If this heat is not sufficient, the forced-air *furnace* will turn on to boost the temperature. If the house thermostat does not call for heat, the solar heated water still travels through the heat exchanger and into the storage tank but the blower fan remains off. On cloudy days or at night when the collectors cannot contribute heat, the zone valves bypass the collectors, circulating heated water from storage through the exchanger, provided that the tank is above 75° F and that room heat is required. If the tank is below this temperature, all heat must be supplied by the furnace. Generally this occurs late at night and early in the morning. On sunny days, the collectors may contribute enough heat from about 9:00 a.m. to 4:00 p.m. to heat the house and increase the temperature of the storage tank. In late afternoon and evening, stored heat is used until the tank drops to 75° F, then the regular furnace takes over.

In *single glazing* areas, an above-ground concrete septic-type storage tank may be considered. The tank must have sufficient strength and be suitably sealed to prevent leaks. Expansion, contraction, settling, and porosity in the concrete all contribute to potential leakage problems, so sealing is essential. If the tank lid is removable, application of a butyl rubber sealing compound on all inside surfaces is suggested. In areas where *dual glazing* is suggested, you will generally find that an above-ground tank located outside loses too much heat at night to be practical. For these conditions, alternatives are locating the storage tank in a basement or burying it. If seepage or leaks would present a problem, I would advise you to consider only a steel or fiberglass tank. Regardless of your choice, consider an *average* storage tank temperature of 95° F. If this value is used in your design, the *amount* of insulation used to insulate the tank will be a function of the temperature difference between the ambient air and storage (95° F) plus the efficiency of the insulating material chosen (see Table 2-9). The *top* of the tank is a most critical area for heat losses. *Wet* insulation is useless; if a *closed-cell* material is not used, provisions must be made to waterproof the assembly.

Using treated water at low pressure as a heat-transfer fluid allows the system piping to be constructed of less expensive CPVC plastic and it allows aluminum absorbers to be considered. The only changes in construction techniques for the *collector* designs I previously suggested would be the substitution of a commercially available aluminum absorber for the copper ones, and the use of plastic materials for glazing instead of glass. Details on connecting the absorber, the CPVC pipe, and the glazing with plastics are included in the sections that follow.

SIZING THE SYSTEM

Selecting the most economical size for your solar space heating system is not difficult once you know the following factors:

- Btus needed to heat the house
- the climate—including solar radiation and ambient temperature
- operating efficiency of the system
- percentage of the heating load that solar energy must provide

If you know your heating load, or can calculate it by analyzing your actual fuel bills, all the better. If not, an *assumption* of the load must be made based on historical averages. One method used by designers is the *heating degree day*, as mentioned earlier. This method assumes that when the outdoor temperature is 65° F or above, no furnace heat will be required to maintain an indoor temperature of 70° F because body heat, lights, cooking, etc., will make up the difference. When house furnaces are installed, they are generally *oversized*. *Design* temperatures are based on a reading of 15° F above the *lowest* temperature ever recorded by the meteorological station in the area. But this low-temperature condition seldom occurs, resulting in unused heating capacity.

A monthly degree day summary for major cities is provided in Table 9-1. Here is an example of how it works. Assume you have a 2500 square foot house in Grand Junction, Colorado, which you have determined uses 900,000 Btus of fuel

Table 9-1. Normal Total Heating Degree Days, Monthly and Annual (Base 65° F).

States and Stations	JULY	AUG	SEPT	OCT	NOV	DEC	JAN	FEB	MAR	APR	MAY	JUNE	ANNUAL
ALABAMA													
Birmingham	0	0	6	93	363	555	592	462	363	108	9	0	2551
Huntsville	0	0	12	127	426	663	694	557	434	138	19	0	3070
Mobile	0	0	0	22	213	357	415	300	211	42	0	0	1560
Montgomery	0	0	0	68	330	527	543	417	316	90	0	0	2291
ALASKA													
Anchorage	245	291	516	930	1284	1572	1631	1316	1293	879	592	315	10864
Annette	242	208	327	567	738	899	949	837	843	648	490	321	7069
Barrow	803	840	1035	1500	1971	2362	2517	2332	2468	1944	1445	957	20174
Barter Is.	735	775	987	1482	1944	2337	2536	2369	2477	1923	1373	924	19862
Bethel	319	394	612	1042	1434	1866	1903	1590	1655	1173	806	402	13196
Cold Bay	474	425	525	772	918	1122	1153	1036	1122	951	791	591	9880
Cordova	366	391	522	781	1017	1221	1299	1086	1113	864	660	444	9764
Fairbanks	171	332	642	1203	1833	2254	2359	1901	1739	1068	555	222	14279
Juneau	301	338	483	725	921	1135	1237	1070	1073	810	601	381	9075
King Salmon	313	322	513	908	1290	1606	1600	1333	1411	966	673	408	11343
Kotzebue	381	446	723	1249	1728	2127	2192	1932	2080	1554	1057	636	16105
McGrath	208	338	633	1184	1791	2232	2294	1817	1758	1122	648	258	14283
Nome	481	496	693	1094	1455	1820	1879	1666	1770	1314	930	573	14171
Saint Paul	605	539	612	862	963	1197	1228	1168	1265	1098	936	726	11199
Shemya	577	475	501	784	876	1042	1045	958	1011	885	837	696	9687
Yakutat	338	347	474	716	936	1144	1169	1019	1042	840	632	435	9092
ARIZONA													
Flagstaff	46	68	201	558	867	1073	1169	991	991	651	437	180	7152
Phoenix	0	0	0	22	234	415	474	328	217	75	0	0	1765
Prescott	0	0	27	245	579	797	865	711	605	360	158	15	4362
Tucson	0	0	0	25	231	406	471	344	242	75	6	0	1800
Winslow	0	0	6	245	711	1008	1054	770	601	291	96	0	4782
Yuma	0	0	0	0	148	319	363	228	130	29	0	0	1217
ARKANSAS													
Fort Smith	0	0	12	127	450	704	781	596	456	144	22	0	3292
Little Rock	0	0	9	127	465	716	756	577	434	126	9	0	3219
Texarkana	0	0	0	78	345	561	626	468	350	105	0	0	2533
CALIFORNIA													
Bakersfield	0	0	0	37	282	502	546	364	267	105	19	0	2122
Bishop	0	0	42	248	576	797	874	666	539	306	143	36	4227
Blue Canyon	34	50	120	347	579	766	865	781	791	582	397	195	5507
Burbank	0	0	6	43	177	301	366	277	239	138	81	18	1646
Eureka	270	257	258	329	414	499	546	470	505	438	372	285	4643
Fresno	0	0	0	78	339	558	586	406	319	150	56	0	2492
Long Beach	0	0	12	40	156	288	375	297	267	168	90	18	1711
Los Angeles	28	22	42	78	180	291	372	302	288	219	158	81	2061
Mt. Shasta	25	34	123	406	696	902	983	784	738	525	347	159	5722
Oakland	53	50	45	127	309	481	527	400	353	255	180	90	2870
Point Arguello	202	186	162	205	291	400	474	392	403	339	298	243	3595
Red Bluff	0	0	0	53	318	555	605	428	341	168	47	0	2515
Sacramento	0	0	12	81	363	577	614	442	360	216	102	6	2773
Sandberg	0	0	30	202	480	691	778	661	620	426	264	57	4209
San Diego	6	0	15	37	123	251	313	249	202	123	84	36	1439
San Francisco	81	78	60	143	306	462	508	395	363	279	214	126	3015
Santa Catalina	16	0	9	50	165	279	353	308	326	249	192	105	2052
Santa Maria	99	93	96	146	270	391	459	370	363	282	233	165	2967
COLORADO													
Alamosa	65	99	279	639	1065	1420	1476	1162	1020	696	440	168	8529
Colo. Springs	9	25	132	456	825	1032	1128	938	893	582	319	84	6423
Denver	6	9	117	428	819	1035	1132	938	887	558	288	66	6283
Grand Junction	0	0	30	313	786	1113	1209	907	729	387	146	21	5641
Pueblo	0	0	54	326	750	986	1085	871	772	429	174	15	5462
CONNECTICUT													
Bridgeport	0	0	66	307	615	986	1079	966	853	510	208	27	5617
Hartford	0	6	99	372	711	1119	1209	1061	899	495	177	24	6172
New Haven	0	12	87	347	648	1011	1097	991	871	543	245	45	5897
DELAWARE													
Wilmington	0	0	51	270	588	927	980	874	735	387	112	6	4930
FLORIDA													
Apalachicola	0	0	0	16	153	319	347	260	180	33	0	0	1308
Daytona Beach	0	0	0	0	75	211	248	190	140	15	0	0	879
Fort Myers	0	0	0	0	24	109	146	101	62	0	0	0	442
Jacksonville	0	0	0	12	144	310	332	246	174	21	0	0	1239
Key West	0	0	0	0	0	28	40	31	9	0	0	0	108
Lakeland	0	0	0	0	57	164	195	146	99	0	0	0	661
Miami Beach	0	0	0	0	0	40	56	36	9	0	0	0	141
Orlando	0	0	0	0	72	198	220	165	105	6	0	0	766
Pensacola	0	0	0	19	195	353	400	277	183	36	0	0	1463
Tallahassee	0	0	0	28	198	360	375	286	202	36	0	0	1485
Tampa	0	0	0	0	60	171	202	148	102	0	0	0	683
West Palm Beach	0	0	0	0	6	65	87	64	31	0	0	0	253
GEORGIA													
Athens	0	0	12	115	405	632	642	529	431	141	22	0	2929
Atlanta	0	0	18	127	414	626	639	529	437	168	25	0	2983
Augusta	0	0	0	78	333	552	549	445	350	90	0	0	2397
Columbus	0	0	0	87	333	543	552	434	338	96	0	0	2383

Table 9-1. Cont'd.

States and Stations	JULY	AUG	SEPT	OCT	NOV	DEC	JAN	FEB	MAR	APR	MAY	JUNE	ANNUAL
Macon	0	0	0	71	297	502	505	403	295	63	0	0	2136
Rome	0	0	24	161	474	701	710	577	468	177	34	0	3326
Savannah	0	0	0	47	246	437	437	353	254	45	0	0	1819
Thomasville	0	0	0	25	198	366	394	305	208	33	0	0	1529
IDAHO													
Boise	0	0	132	415	792	1017	1113	854	722	438	245	81	5809
Idaho Falls 46W	16	34	270	623	1056	1370	1538	1249	1085	651	391	192	8475
Idaho Falls 42 NW	16	40	282	648	1107	1432	1600	1291	1107	657	388	192	8760
Lewiston	0	0	123	403	756	933	1063	815	694	426	239	90	5542
Pocatello	0	0	172	493	900	1166	1324	1058	905	555	319	141	7033
ILLINOIS													
Cairo	0	0	36	164	513	791	856	680	539	195	47	0	3821
Chicago	0	0	81	326	753	1113	1209	1044	890	480	211	48	6155
Moline	0	9	99	335	774	1181	1314	1100	918	450	189	39	6408
Peoria	0	6	87	326	759	1113	1218	1025	849	426	183	33	6025
Rockford	6	9	114	400	837	1221	1333	1137	961	516	236	60	6830
Springfield	0	0	72	291	696	1023	1135	935	769	354	136	18	5429
INDIANA													
Evansville	0	0	66	220	606	896	955	767	620	237	68	0	4435
Fort Wayne	0	9	105	378	783	1135	1178	1028	890	471	189	39	6205
Indianapolis	0	0	90	316	723	1051	1113	949	809	432	177	39	5699
South Bend	0	6	111	372	777	1125	1221	1070	933	525	239	60	6439
IOWA													
Burlington	0	0	93	322	768	1135	1259	1042	859	426	177	33	6114
Des Moines	0	9	99	363	837	1231	1398	1165	967	489	211	39	6808
Dubuque	12	31	156	450	906	1287	1420	1204	1026	546	260	78	7376
Sioux City	0	9	108	369	867	1240	1435	1198	989	483	214	39	6951
Waterloo	12	19	138	428	909	1296	1460	1221	1023	531	229	54	7320
KANSAS													
Concordia	0	0	57	276	705	1023	1163	935	781	372	149	18	5479
Dodge City	0	0	33	251	666	939	1051	840	719	354	124	9	4986
Goodland	0	6	81	381	810	1073	1166	955	884	507	236	42	6141
Topeka	0	0	57	270	672	980	1122	893	722	330	124	12	5182
Wichita	0	0	33	229	618	905	1023	804	645	270	87	6	4620
KENTUCKY													
Covington	0	0	75	291	669	983	1035	893	756	390	149	24	5265
Lexington	0	0	54	239	609	902	946	818	685	325	105	0	4683
Louisville	0	0	54	248	609	890	930	818	682	315	105	9	4660
LOUISIANA													
Alexandria	0	0	0	56	273	431	471	361	260	69	0	0	1921
Baton Rouge	0	0	0	31	216	369	409	294	208	33	0	0	1560
Burrwood	0	0	0	0	96	214	298	218	171	27	0	0	1024
Lake Charles	0	0	0	19	210	341	381	274	195	39	0	0	1459
New Orleans	0	0	0	19	192	322	363	258	192	39	0	0	1385
Shreveport	0	0	0	47	297	477	552	426	304	81	0	0	2184
MAINE													
Caribou	78	115	336	682	1044	1535	1690	1470	1308	858	468	183	9767
Portland	12	53	195	508	807	1215	1339	1182	1042	675	372	111	7511
MARYLAND													
Baltimore	0	0	48	264	585	905	936	820	679	327	90	0	4654
Frederick	0	0	66	307	624	955	995	876	741	384	127	12	5087
MASSACHUSETTS													
Blue Hill Obs.	0	22	108	381	690	1085	1178	1053	936	579	267	69	6368
Boston	0	9	60	316	603	983	1088	972	846	513	208	36	5634
Nantucket	12	22	93	332	573	896	992	941	896	621	384	129	5891
Pittsfield	25	59	219	524	831	1231	1339	1196	1063	660	326	105	7578
Worcester	6	34	147	450	774	1172	1271	1123	998	612	304	78	6969
MICHIGAN													
Alpena	68	105	273	580	912	1268	1404	1299	1218	777	446	156	8506
Detroit (city)	0	0	87	360	738	1088	1181	1058	936	522	220	42	6232
Escanaba	59	87	243	539	924	1293	1445	1296	1203	777	456	159	8481
Flint	16	40	159	465	843	1212	1330	1198	1066	639	319	90	7377
Grand Rapids	9	28	135	434	804	1147	1259	1134	1011	579	279	75	6894
Lansing	6	22	138	431	813	1163	1262	1142	1011	579	273	69	6909
Marquette	59	81	240	527	936	1268	1411	1268	1187	771	468	177	8393
Muskegon	12	28	120	400	762	1088	1209	1100	995	594	310	78	6696
Sault Ste. Marie	96	105	279	580	951	1367	1525	1380	1277	810	477	201	9048
MINNESOTA													
Duluth	71	109	330	632	1131	1581	1745	1518	1355	840	490	198	10000
Internat'l Falls	71	112	363	701	1236	1724	1919	1621	1414	828	443	174	10606
Minneapolis	22	31	189	505	1014	1454	1631	1380	1166	621	288	81	8382
Rochester	25	34	186	474	1005	1438	1593	1366	1150	630	301	93	8295
Saint Cloud	28	47	225	549	1065	1500	1702	1445	1221	666	326	105	8879
MISSISSIPPI													
Jackson	0	0	0	65	315	502	546	414	310	87	0	0	2239
Meridian	0	0	0	81	339	518	543	417	310	81	0	0	2289
Vicksburg	0	0	0	53	279	462	512	384	282	69	0	0	2041
MISSOURI													
Columbia	0	0	54	251	651	967	1076	874	716	324	121	12	5046
Kansas	0	0	39	220	612	905	1032	818	682	294	109	0	4711

Table 9-1. Cont'd.

States and Stations	JULY	AUG	SEPT	OCT	NOV	DEC	JAN	FEB	MAR	APR	MAY	JUNE	ANNUAL
St. Joseph	0	6	60	285	708	1039	1172	949	769	348	133	15	5484
St. Louis	0	0	60	251	627	936	1026	848	704	312	121	15	4900
Springfield	0	0	45	223	600	877	973	781	660	291	105	6	4561
MONTANA													
Billings	6	15	186	487	897	1135	1296	1100	970	570	285	102	7049
Glasgow	31	47	270	608	1104	1466	1711	1439	1187	648	335	150	8996
Great Falls	28	53	258	543	921	1169	1349	1154	1063	642	384	186	7750
Havre	28	53	306	595	1065	1367	1584	1364	1181	657	338	162	8700
Helena	31	59	294	601	1002	1265	1438	1170	1042	651	381	195	8129
Kalispell	50	99	321	654	1020	1240	1401	1134	1029	639	397	207	8191
Miles City	6	6	174	502	972	1296	1504	1252	1057	579	276	99	7723
Missoula	34	74	303	651	1035	1287	1420	1120	970	621	391	219	8125
NEBRASKA													
Grand Island	0	6	108	381	834	1172	1314	1089	908	462	211	45	6530
Lincoln	0	6	75	301	726	1066	1237	1016	834	402	171	30	5864
Norfolk	9	0	111	397	873	1234	1414	1179	983	498	233	48	6979
North Platte	0	6	123	440	885	1166	1271	1039	930	519	248	57	6684
Omaha	0	12	105	357	828	1175	1355	1126	939	465	208	42	6612
Scottsbluff	0	0	138	459	876	1128	1231	1008	921	552	285	75	6673
Valentine	9	12	165	493	942	1237	1395	1176	1045	579	288	84	7425
NEVADA													
Elko	9	34	225	561	924	1197	1314	1036	911	621	409	192	7433
Ely	28	43	234	592	939	1184	1308	1075	977	672	456	225	7733
Las Vegas	0	0	0	78	387	617	688	487	335	111	6	0	2709
Reno	43	87	204	490	801	1026	1073	823	729	510	357	189	6332
Winnemucca	0	34	210	536	876	1091	1172	916	837	573	363	153	6761
HEW HAMPSHIRE													
Concord	6	50	177	505	822	1240	1358	1184	1032	636	298	75	7383
Mt. Wash. Obs.	493	536	720	1057	1341	1742	1820	1663	1652	1260	930	603	13817
NEW JERSEY													
Atlantic City	0	0	39	251	549	880	936	848	741	420	133	15	4812
Newark	0	0	30	248	573	921	983	876	729	381	118	0	4859
Trenton	0	0	57	264	576	924	989	885	753	399	121	12	4980
NEW MEXICO													
Albuquerque	0	0	12	229	642	868	930	703	595	288	81	0	4348
Clayton	0	6	66	310	699	899	986	812	747	429	183	21	5158
Raton	9	28	126	431	825	1048	1116	904	834	543	301	63	6228
Roswell	0	0	18	202	573	806	840	641	481	201	31	0	3793
Silver City	0	0	6	183	525	729	791	605	518	261	87	0	3705
NEW YORK													
Albany	0	19	138	440	777	1194	1311	1156	992	564	239	45	6875
Binghamton (AP)	22	65	201	471	810	1184	1277	1154	1045	645	313	99	7286
Binghamton (PO)	0	28	141	406	732	1107	1190	1081	949	543	229	45	6451
Buffalo	19	37	141	440	777	1156	1256	1145	1039	645	329	78	7062
Central Park	0	0	30	233	540	902	986	885	760	408	118	9	4871
JFK International	0	0	36	248	564	933	1029	935	815	480	167	12	5219
La Buardia	0	0	27	223	528	887	973	879	750	414	124	6	4811
Rochester	9	31	126	415	747	1125	1234	1123	1014	597	279	48	6748
Schenectady	0	22	123	422	756	1159	1283	1131	970	543	211	30	6650
Syracuse	6	28	132	415	744	1153	1271	1140	1004	570	248	45	6756
NORTH CAROLINA													
Asheville	0	0	48	245	555	775	784	683	592	273	87	0	4042
Cape Hatteras	0	0	0	78	273	521	580	518	440	177	25	0	2612
Charlotte	0	0	6	124	438	691	691	582	481	156	22	0	3191
Greensboro	0	0	33	192	513	778	784	672	552	234	47	0	3805
Raleigh	0	0	21	164	450	716	725	616	487	180	34	0	3393
Wilmington	0	0	0	74	291	521	546	462	357	96	0	0	2347
Winston Salem	0	0	21	171	483	474	753	652	524	207	37	0	3595
NORTH DAKOTA													
Bismarck	34	28	222	577	1083	1463	1708	1442	1203	645	329	117	8851
Devils Lake	40	53	273	642	1191	1634	1872	1579	1345	753	381	138	9901
Fargo	28	37	219	574	1107	1569	1789	1520	1262	690	332	99	9226
Williston	31	43	261	601	1122	1513	1758	1473	1262	681	357	141	9243
OHIO													
Akron	0	9	96	381	726	1070	1138	1016	871	489	202	39	6037
Cincinnati	0	0	54	248	612	921	970	837	701	336	118	9	4806
Cleveland	9	25	105	384	738	1088	1159	1047	918	552	260	66	6351
Columbus	0	6	84	347	714	1039	1088	949	809	426	171	27	5660
Dayton	0	6	78	310	696	1045	1097	955	809	429	167	30	5622
Mansfield	9	22	114	397	768	1110	1169	1042	924	543	245	60	6403
Sandusky	0	6	66	313	684	1032	1107	991	868	495	198	36	5796
Toledo	0	16	117	406	792	1138	1200	1056	924	543	242	60	6494
Youngstown	0	19	120	412	771	1104	1169	1047	921	540	248	60	6417
OKLAHOMA													
Okla. City	0	0	15	164	498	766	868	664	527	189	34	0	3725
Tulsa	0	0	18	158	522	787	893	683	539	213	47	0	3860
OREGON													
Astoria	146	130	210	375	561	679	753	622	636	480	363	231	5186
Burns	12	37	210	515	867	1113	1246	988	856	570	366	177	6957
Eugene	34	34	129	366	585	719	803	627	589	426	279	135	4726
Meacham	84	124	288	580	918	1091	1209	1005	983	726	527	339	7874
Medford	0	0	78	372	678	871	918	697	642	432	242	78	5008
Pendleton	0	0	111	350	711	884	1017	773	617	396	205	63	5127
Portland	25	28	114	335	597	735	825	644	586	396	245	105	4635

Table 9-1. Cont'd.

States and Stations	JULY	AUG	SEPT	OCT	NOV	DEC	JAN	FEB	MAR	APR	MAY	JUNE	ANNUAL
Roseburg	22	16	105	329	567	713	766	608	570	405	267	123	4491
Salem	37	31	111	338	594	729	822	647	611	417	273	144	4754
Sexton Summit	81	81	171	443	666	874	958	809	818	609	465	279	6254
PENNSYLVANIA													
Allentown	0	0	90	353	693	1045	1116	1002	849	471	167	24	5810
Erie	0	25	102	391	714	1063	1169	1081	973	585	288	60	6451
Harrisburg	0	0	63	298	648	992	1045	907	766	396	124	12	5251
Philadelphia	0	0	60	291	621	964	1014	890	744	390	115	12	5101
Pittsburgh	0	9	105	375	726	1063	1119	1002	874	480	195	39	5987
Reading	0	0	54	257	597	939	1001	885	735	372	105	0	4945
Scranton	0	19	132	434	762	1104	1156	1028	893	498	195	33	6254
Williamsport	0	9	111	375	717	1073	1122	1002	856	468	177	24	5934
RHODE ISLAND													
Block Is.	0	16	78	307	594	902	1020	955	877	612	344	99	5804
Providence	0	16	96	372	660	1023	1110	988	868	534	236	51	5954
SOUTH CAROLINA													
Charleston	0	0	0	59	282	471	487	389	291	54	0	0	2033
Columbia	0	0	0	84	345	577	570	470	357	81	0	0	2484
Florence	0	0	0	78	315	552	552	459	347	84	0	0	2387
Greenville	0	0	0	112	387	636	648	535	434	120	12	0	2884
Spartanburg	0	0	15	130	417	667	663	560	453	144	25	0	3074
SOUTH DAKOTA													
Huron	9	12	165	508	1014	1432	1628	1355	1125	600	288	87	8223
Rapid City	22	12	165	481	897	1172	1333	1145	1051	615	326	126	7345
Sioux Falls	19	25	168	462	972	1361	1544	1285	1082	573	270	78	7839
TENNESSEE													
Bristol	0	0	51	236	573	828	828	700	598	261	68	0	4143
Chattanooga	0	0	18	143	468	698	722	577	453	150	25	0	3254
Knoxville	0	0	30	171	489	725	732	613	493	198	43	0	3494
Memphis	0	0	18	130	447	698	729	585	456	147	22	0	3232
Nashville	0	0	30	158	495	732	778	644	512	189	40	0	3578
Oak Ridge (CO)	0	0	39	192	531	772	778	669	552	228	56	0	3817
TEXAS													
Abilene	0	0	0	99	366	586	642	470	347	114	0	0	2624
Amarillo	0	0	18	205	570	797	877	664	546	252	56	0	3985
Austin	0	0	0	31	225	388	468	325	223	51	0	0	1711
Brownsville	0	0	0	0	66	149	205	106	74	0	0	0	600
Corpus Christi	0	0	0	0	120	220	291	174	109	0	0	0	914
Dallas	0	0	0	62	321	524	601	440	319	90	6	0	2363
El Paso	0	0	0	84	414	648	685	445	319	105	0	0	2700
Forth Worth	0	0	0	65	324	536	614	448	319	99	0	0	2405
Galveston	0	0	0	0	138	270	350	258	189	30	0	0	1235
Houston	0	0	0	6	183	307	384	288	192	36	0	0	1396
Laredo	0	0	0	0	105	217	267	134	74	0	0	0	797
Lubbock	0	0	18	174	513	744	800	613	484	201	31	0	3578
Midland	0	0	0	87	381	592	651	468	322	90	0	0	2591
Port Arthur	0	0	0	22	207	329	384	274	192	39	0	0	1447
San Angelo	0	0	0	68	318	536	567	412	288	66	0	0	2255
San Antonio	0	0	0	31	207	363	428	286	195	39	0	0	1549
Victoria	0	0	0	6	150	270	344	230	152	21	0	0	1173
Waco	0	0	0	43	270	456	536	389	270	66	0	0	2030
Wichita Falls	0	0	0	99	381	632	698	518	378	120	6	0	2832
UTAH													
Milford	0	0	99	443	867	1141	1252	988	822	519	279	87	6497
Salt Lake City	0	0	81	419	849	1082	1172	910	763	459	233	84	6052
Wendover	0	0	40	372	822	1091	1178	902	729	408	177	51	5778
VERMONT													
Burlington	28	65	207	539	891	1349	1513	1333	1187	714	353	90	8269
VIRGINIA													
Cape Henry	0	0	0	112	360	645	694	633	536	246	53	0	3279
Lynchburg	0	0	51	223	540	822	849	731	605	267	78	0	4166
Norfolk	0	0	0	136	408	698	738	655	533	216	37	0	3421
Richmond	0	0	36	214	495	784	815	703	546	219	53	0	3865
Roanoke	0	0	51	229	549	825	834	722	614	261	65	0	4150
Wash. Nat'l AP	0	0	33	217	519	834	871	762	626	288	74	0	4224
WASHINGTON													
Olympia	68	71	198	422	636	753	834	675	645	450	307	177	5236
Seattle	50	47	129	329	543	657	738	599	577	396	242	117	4424
Seattle Boeing	34	40	147	384	624	763	831	655	608	411	242	99	4838
Seattle Tacoma	56	62	162	391	633	750	828	678	657	474	295	159	5145
Spokane	9	25	168	493	879	1082	1231	980	834	531	288	135	6655
Stampede Pass	273	291	393	701	1008	1178	1287	1075	1085	855	654	483	9283
Tatoosh Is.	295	279	306	406	534	639	713	613	645	525	431	333	5719
Walla Walla	0	0	87	310	681	834	986	745	589	342	177	45	4805
Yakima	0	12	144	450	828	1039	1163	868	713	435	220	69	5941
WEST VIRGINIA													
Charleston	0	0	63	254	591	865	880	770	648	300	96	9	4476
Elkins	9	25	135	400	729	992	1008	896	791	444	198	48	5675
Huntington	0	0	63	257	585	856	880	764	636	294	99	12	4446
Parkersburg	0	0	60	264	606	905	942	826	691	339	115	6	4754
WISCONSIN													
Green Bay	28	50	174	484	924	1333	1494	1313	1141	654	335	99	8029
LaCrosse	12	19	153	437	924	1339	1504	1277	1070	540	245	69	7589
Madison	25	40	174	474	930	1330	1473	1274	1113	618	310	102	7863
Milwaukee	43	47	174	471	876	1252	1376	1193	1054	642	372	135	7635
WYOMING													
Casper	6	16	192	524	942	1169	1290	1084	1020	657	381	129	7410
Cheyenne	19	31	210	543	924	1101	1228	1056	1011	672	381	102	7278
Lander	6	19	204	555	1020	1299	1417	1145	1017	654	381	153	7870
Sheridan	25	31	219	539	948	1200	1355	1154	1054	642	366	150	7683

Climatic Atlas of the United States. U.S. Department of Commerce per day during the month of December based on actual fuel bills (assume a gas furnace is 60 percent efficient). Looking at the table, Grand Junction typically has 1113 degree days in December, or 36 (1113/31) degree days in a 24-hour period. You can now calculate the number of Btus per degree day per square foot that your home requires:

$$\frac{900,000}{36 \times 2500} = 10 \text{ Btus per degree day per square foot}$$

The average American home is *not* well insulated and would have a reading in the range of 12 to 20 Btus per degree day per square foot. *Before considering a solar heater, you should insulate as necessary to bring your home into the 8 to 12 range.*

Collector Area

The Los Alamos Scientific Laboratory in New Mexico has conducted an hour-by-hour computer simulation of a *standard* liquid solar heating system to determine the ratio of building load to solar collector area for 85 key cities. These data are shown in Table 9-2. To demonstrate the use of the table, the 2500 square foot house in Grand Junction, Colorado, with a thermal load of 10 Btus per degree day per square foot will have a total building load of 25,000 (2500 × 10) Btus per degree day. If you want the solar heater to provide 75 percent of the heating load, Table 9-2 gives a value of 22 Btus per degree day per square foot of collector. The required collector area, therefore, is

$$\frac{25,000}{22} = 1136 \text{ square feet}$$

For 50 percent of the load the table shows Grand Junction to have 46 Btus per degree day per square foot of collector, so:

$$\frac{25,000}{46} = 543 \text{ square feet of collectors needed}$$

These figures are not exact, as there are differences between the *standard* liquid system defined by Los Alamos and the one in this chapter; but the figures can be used to give you an approximation of the size of collector needed to maintain the building temperature at 68° F.

Table 9-2. Values of Load to Collector Ratios (LC) for 85 Selected Cities.

LC, BTU per degree day per square foot where solar provides 25%, 50%, 75% of total heat

City	25%	50%	75%
Albuquerque, NM	161	64	31
Apalachicola, FL	324	129	65
Astoria, OR	127	45	19
Atlanta, GA	154	59	29
Bismarck, ND	78	29	14
Blue Hill, MA	82	31	15
Boise, ID	108	39	17
Boston, MA	86	33	16
Brownsville, TX	517	218	110
Burlington, VT	63	24	11
Cape Hatteras, NC	189	74	36
Caribou, ME	68	26	12
Charleston, SC	210	82	41
Cleveland, OH	71	26	12
Columbia, MO	102	38	18
Columbus, OH	77	29	13
Corvallis, OR	120	42	18
Davis, CA	198	72	33
Dodge City, KA	126	49	24
East Lansing, MI	76	28	13
East Wareham, MA	97	37	18
El Centro, CA	547	206	97
El Paso, TX	228	88	44
Ely, NV	119	47	23
Flaming Gorge, UT	111	43	21
Forth Worth, TX	186	73	37
Fresno, CA	195	70	32
Glasgow, MT	105	41	20
Granby, CO	119	47	23
Grand Junction CO	119	46	22
Great Falls, MT	93	35	16
Greensboro, NC	128	50	24
Griffin, GA	217	84	42
Hattaras, NC	204	79	39
Indianapolis, IN	86	32	15
Inyokern, CA	232	88	42
Ithaca, NY	68	24	11
Lake Charles, LA (el. 39 ft)	261	104	53
(el. 60 ft)	244	96	48
Lander, WY	108	42	21
Laramie, WY	106	42	21
Las Vegas, NV	218	84	42
Lemont, IL	79	30	14
Lincoln, NE	104	39	19
Little Rock, AR	126	48	24
Los Alamos, NM	107	41	21
Los Angeles, CA	416	157	75
Madison, WI	76	28	13
Medford, OR	107	38	16
Midland, TX	202	79	39
Nashville, TN	117	44	21
Newport, RI	97	37	18
New York, NY	88	34	16
North Omaha, NE	89	34	16
Oak Ridge, TN	111	42	20
Oklahoma City, OK	134	53	26
Page, AR	128	48	23
Phoenix, AR	300	118	59
Prosser, WA	117	41	18
Pullman, WA	100	36	16
Put-In-Bay, OH	68	24	11
Raleigh, NC	133	52	25
Rapid City, SD	97	37	18
Reno, NV	125	47	22
Richland, WA	100	35	15
Riverside, CA	391	152	74

Table 9-2. Cont'd.

City	25%	50%	75%
Saint Cloud, NM	71	27	13
Salt Lake City, UT	107	40	19
San Antonio, TX	262	103	52
Santa Maria, CA	353	142	67
Sault Ste. Marie, MI	74	27	12
Sayville, NY	98	38	18
Seabrook, NJ	97	37	18
Seattle, WA	94	33	13
Schenectady, NY	63	24	11
Shreveport, VA	179	70	35
Silver Hill, MD	111	43	21
Spokane, WA	90	31	14
State College, PA	78	29	14
Sterling, VA	111	43	21
Stillwater, OK	132	52	25
Tallahassee, FL	283	113	57
Toronto, Canada	72	27	13
Tuscon, AR	301	118	59
Winnipeg, Canada	63	23	11

Source·

Los Alamos Scientific Laboratory
Los Alamos, New Mexico 87545

Storage Tank Size

You should plan to store 1.5 gallons of water for each square foot of collector in addition to an allowance for 1 gallon expansion at the top of the tank for each 50 gallons stored. Storage tanks come in standard sizes, so you may end up with one slightly larger than you actually need for the heating load.

Heat Exchanger

You will need to size the heat exchanger for each installation. As mentioned in the system description, heated water returning from the collectors always circulates through the water-to-air heat exchanger in the furnace plenum. Not only is it necessary to select an exchanger with a *physical* size that can be assembled to your heater, but the *capacity* of the exchanger must be correct to transfer the desired amount of heat. Start by referring to the nameplate on your existing furnace and blower. Record the capacity of the furnace in Btus and the capacity of the fan in cubic feet per minutes (cfm). The exchanger should be of sufficient size to transfer one-fourth of the rated output of the furnace from the solar heated water to the air blowing across the exchanger coils. The flow rate of water through the exchanger should be set at 0.8 gallons per minute (gpm) for each 100 square feet of collectors specified. Other specifications are:

- air friction *pressure drop* across exchanger not to exceed 0.1 inch of water
- water inlet temperature of 105° F
- air inlet temperature of 68° F
- air outlet temperature of 90° F

An experienced HVAC contractor or heat exchanger manufacturer can help you in your selection.

Pump Capacity

A single pump is used in this system. It must have the capacity to refill the collector lines after they have drained and to maintain the desired flow rate through the collectors and heat exchanger. As discussed in Chapter 7, the pump load is at a maximum as the collector lines are being filled with water. Once filled, a siphoning action occurs, reducing the load. You will need to calculate the total *head* in feet for your specific system. The *primary* head loss will be the vertical rise from the pump to the top of the collector assembly, but the effects of *all* components between the pump and air vent valve should be added to determine the total head. This includes line loss and pressure drops across collectors and one of the 3-way valves.

Now refer to the performance curve for the pump selected, to insure adequate capacity. The design flow rate through the system is 0.8 gpm for each 100 square feet of collectors. The pump selection then depends upon the head and system flow for your specific installation. I recommend that you add a throttling valve in the collector feed line, adjusting the flow to the desired level once the system is operating.

MATERIALS: AVAILABILITY AND COST

The materials that follow cost approximately $4180 as of this writing. (Chapters 2 and 3 provide detailed information should you wish to make substitutions.) No return or feed lines are included in the estimate, because the length and complexity of these lines will vary from job to job. The collector specifications are for a self-built unit with a material cost of approximately $5 per square foot of *effective* absorber area.

Collector design

number of covers....................1
cover material................4 mil thick Tedlar PVF film type 400 BG 20 TR

absorber material ....aluminum Roll-Bond
absorber coating ....flat-black paint
insulation ....industrial grade fiberglass
collector frame ....extruded aluminum
nominal size ....36″ × 98″ × 4″

Make your own estimate of material cost from local sources of supply. You can then decide whether it is advantageous to buy a factory-made collector or build your own. Also, when the entire bill of materials is priced out locally, the economic advantages of proceeding with construction of a solar room heating system will be more clearly defined.

Bill of Materials for a 20-collector system with approximately 453 square feet of effective area

Quantity	Size	Description
Absorber:		
20 pcs.	33 3/4″ × 96″	aluminum Roll-Bond, parallel water flow 3/4 in. O.D. inlet and outlet fittings. Olin Brass model FS-7767 (also identified as part number FS-0606) or equal
Collector frame:		
40 pcs.	12 ft	aluminum frame extrusion, Alcoa die number 459831
40 pcs.	12 ft	aluminum glazing cap extrusion, Alcoa die number 459841
1 pc.	12 ft	aluminum hold-down clip extrusion, Alcoa die number 459851
10 gross	No. 10 × 3/4 in.	No. 10 type A, round or pan head sheet-metal screws, 3/4 in. long, cadmium plated steel
15 pcs.	12 ft	aluminum channel extrusion 1/2″ × 1/2″ × 1/16″ (spreader bar—3 required per collector)

Quantity	Size	Description
Back plate:		
20 pcs.	1/4″ × 4′ × 8′	exterior grade or marine plywood, 1/4 in. thickness
Collector insulation:		
170 pcs.	1″ × 24″ × 48″	industrial grade semi-rigid fiberglass board. Certain-Teed Corp. No. 850, Johns-Manville No. 814 or equal, no facings
Glazing:		
170 lin ft	38″ min width	PVF film, 4 mil thickness, DuPont Tedlar type 400 BG 20 TR or equal
10 rolls	1/4″ × 3/8″ × 50′	General Electric Silglaze tape SCT 1535 or equal
4	1/12 gal cartridge	General Electric Silglaze silicone sealant cartridge SCS 1803 or equal
Valves:	*Size Symbol**	
1	3/4″ V – 1	3-way fan coil valve. 125 psi operating pressure. 24 VAC 60 cycle control. Amp rating 0.9. 3/4 in. brass sweat connections. Taco, Inc. model 561 or equal
1	1″ V – 2	3-way fan coil valve. 125 psi operating pressure. 24 VAC 60 cycle control. Amp rating 0.9. 1 in. brass sweat connections. Taco, Inc. model 562 or equal
3	3/4″ V – 3, V – 4, V – 5	swing check valves. Cast brass with 3/4 in. NPT threads both ends
1	3/4″ V – 6	air vent and vacuum breaker. 3/4 in. male NPT threaded connector. Whirlpool Corp. or equal

*Symbols refer to plumbing and electrical schematic drawings, Figs. 9-2 and 9-3.

Quantity	Size	Description
1	3/4″ × 3/4″ V − 7	pressure-relief valve. 50 psi pressure setting, 3/4 ″ NPT threads. Robertshaw Controls, Tempstat model P500 or equal
1	1 1/2 V − 8	balancing valve—full flow design. 150 psi rating, 1 1/2 in. NPT threads both ends. Brass body

Control system:

Size	*Symbol**	
1	DTFC	differential temperature flow controller. 120 VAC output and inputs. Low level outputs for solar panel and storage tank sensors. Pump turn-on at 10° F differential, turn-off at 5° F differential. Deko-Labs model TC-3 or equal. 2 temperature sensors included
1	S − 1	thermostat switch. Bimetallic type including well. Single-pole, single throw, 115 VAC 10 amp rating. Operating temperature 40° F to 200° F. Mercoid type FM437-3-3516 or equal
1	S − 2	house thermostat. Chromalox WR-1E30 or equal
1	T − 1	step down transformer. 120 to 24 VAC, 4 amp. Allied Electronics Corp. type 6K80VBR or equal

*Symbols refer to plumbing and electrical schematic drawings, Figs. 9-2 and 9-3.

Quantity	Size	Description
1	T – 2	step down transformer. 120 to 24 VAC, 1 amp. Allied Electronics Corp. type 6K113HF or equal
2	R – 1 R – 2	relays. 24 VAC, 10 amp. Double-pole, double throw. Allied Electronics Corp. type KA11AG or equal
Pump:		rotary vane, positive displacement type pump (capable of pumping 240 gal/hr of water, internal relief valve must be set at 50 psi, 1/3 hp, 120 volt 60 cycle motor). Procon model C02057HFEP or equal
Heat exchanger:		
1		water-to-air type finned tube heat exchanger without turbulators. Single row coil, type T with specifications as noted. Sized to meet specific application. Trane Co. or equal
Storage tank:		
1		concrete (septic-type), steel or fiberglass as preferred. Sized per specifications in text
Miscellaneous parts and plumbing:		
1	gal	primer for aluminum absorbers. DeSoto Koropon or equal
2	gal	flat-black urethane resin applied over primer. For aluminum absorbers. DeSoto Enersorb or equal
1	1″	pipe strainer for suction side of pump. 1 in. NPT threads, brass body. Strainer to be

Quantity	Size	Description
		fitted with 100 mesh stainless steel screen
as needed		storage tank insulation as necessary to meet specific climatic requirements
20	3/4" I.D.	CPVC thermoplastic couplings. Unthreaded schedule 80. (To insulate absorber inlet and outlet fittings from collector frame. This part is cut into 2 equal pieces)
as needed	3/4" I.D.	high temperature silicone rubber pressure hose
as needed		3/4" extended tang, screw-type hose clamps

COLLECTOR CONSTRUCTION

The collector frame and back plate assembly is similar to details covered in previous chapters. Only two changes should be noted. If you decide to use an aluminum Roll-Bond absorber, it comes with connector tubes welded in place. Model FS-0606 (7767) has the dimensions shown in Fig. 9-2. It will be necessary to leave one end of the frame unassembled until the absorber is positioned due to the overall length of the part. The second change from previous assemblies relates to the glazing method if a PVF film is used instead of glass. In this case, the upper leg of the frame is *not* cut off. It is used to space the film well above the absorber to minimize contact with the hot absorber. Refer to Fig. 9-3 which shows a typical glazing detail. The PVF film will actually shrink the first time it is heated, so provision should be made to support the frame along its 98 inch length with spacer bars of the type shown in Fig. 6-9. The PVF film should be installed when the outdoor temperature is above 70° F to minimize sag.

As in previous designs, CPVC unthreaded couplings are used to insulate the absorber inlet and outlet fittings from the collector frame.

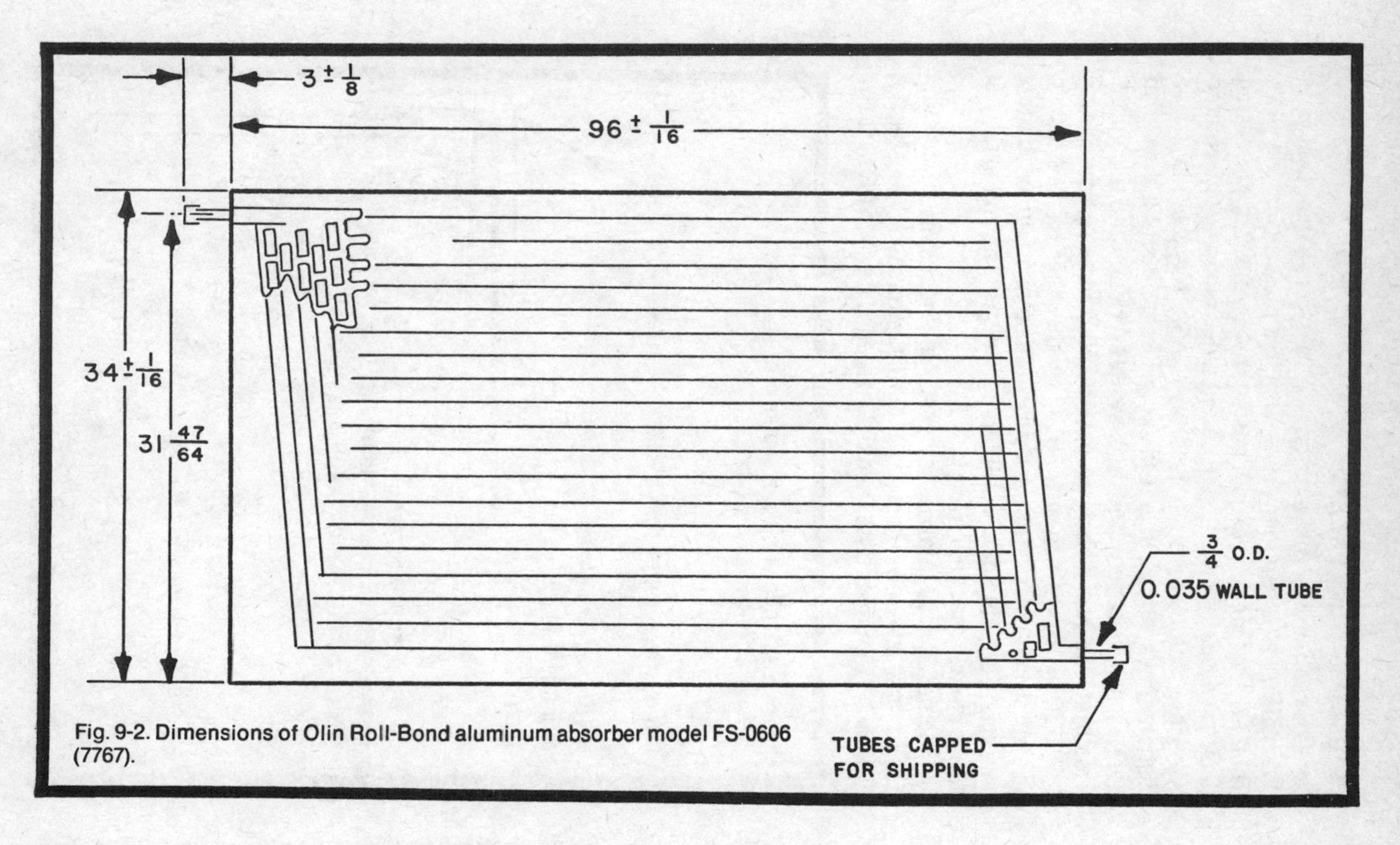

Fig. 9-2. Dimensions of Olin Roll-Bond aluminum absorber model FS-0606 (7767).

PLUMBING THE SYSTEM

Connecting all the elements of a typical installation is straightforward. Follow the schematic diagram of Fig. 9-4. This arrangement was suggested by NASA for its tested demonstration system. CPVC plastic piping (chlorinated polyvinyl chloride) was used by NASA for the hot water distribution system. It is less expensive than metal, has good corrosion resistance, and is easy to fabricate and assemble. This material has a resistance to continuous temperatures up to 180° F at 100 psi pressure. PVC, on the other hand, starts to fail at much lower temperatures, making it an unsuitable material for this solar application. Be sure the material you use is clearly marked "CPVC".

Some tips to making solvent-weld joints in connecting CPVC are provided in a free manual from NIBCO, Inc. of Elkhart, Indiana. The booklet, titled *The NIBCO Plastic*

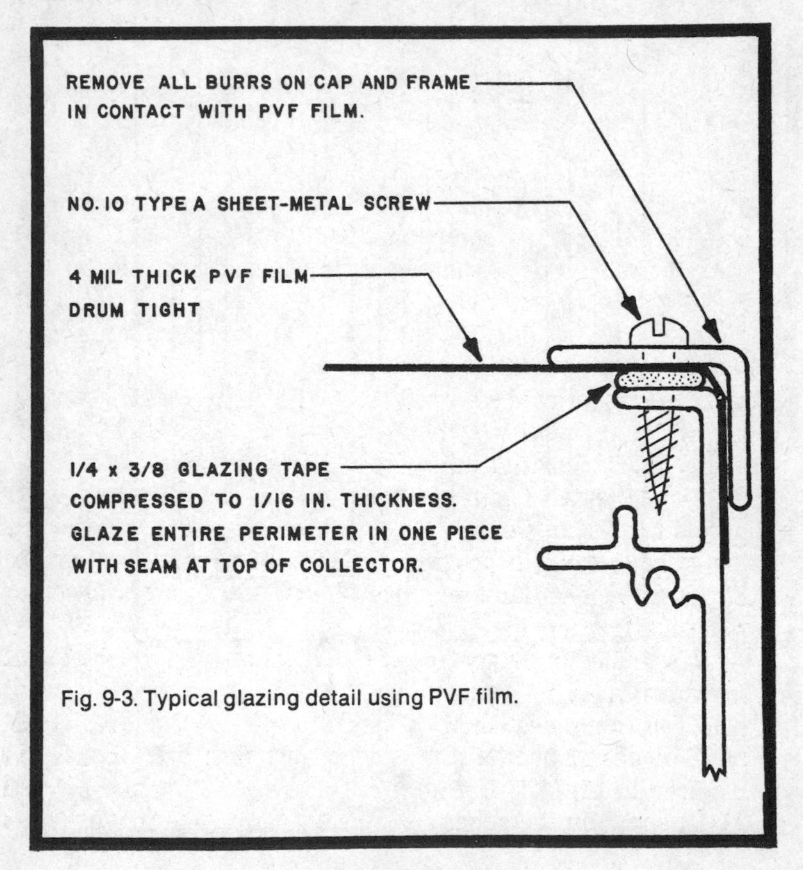

Fig. 9-3. Typical glazing detail using PVF film.

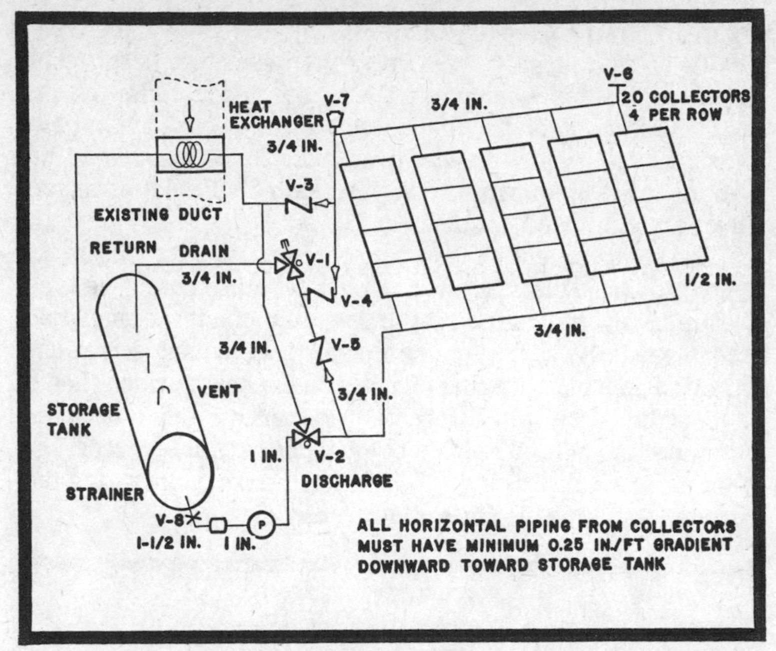

Fig. 9-4. Typical plumbing schematic similar to NASA report N76-27671.

Piping How-to-and-why Pocket Handbook outlines eight steps in making a proper joint. Methods of connecting both sweat and threaded copper components to CPVC are also covered in the handbook. Special transition unions will be required to connect the 3-way valves, heat exchanger, and other metal-to-plastic joints. Solvent-welds are impossible to make with reliability, and although threaded connections may be satisfactory for *cold* water lines, the different expansion rates of metal and plastic may cause leakage if you use them for *hot* lines. CPVC transition unions have proven to be reliable.

Note that all horizontal piping from the collectors must have a minimum slope of 0.25 inch per foot *downward* toward the storage tank to provide drainage and freeze protection. Suggested pipe diameters are shown to promote flow and minimize friction losses.

If you use the 3-way fan coil valves manufactured by Taco, Inc. for valves V−1 and V−2, the following explanation will help you to connect them properly. Both models 561 and 562 have three connections labeled *main*, *unit*, and *bypass*, as pictured in Fig. 9-5. For simplicity, these will be abbreviated M, U, and B in the schematics that follow. Note that a cap is sweat soldered to the *unit* side of valve V–1 to effectively provide an inexpensive drain valve that would be energized in the *closed* position when valve V–2 is energized. Refer to Fig. 9-6. The *heavy* black line indicates the water circulation path when the valves are de-energized. The *light* lines would be used for the drain-down circuit. The path of circulation as the valves are energized, is shown in Fig. 9-7.

The solar collectors are connected using extended tank, screw-type hose clamps and high temperature hose. The type of clamps recommended by the hose manufacturer should be used. Generally, wire-type clamps or those having serrations or slotted openings bearing directly on the hose surface should be *avoided*. Due to differential expansion rates between aluminum, steel, and rubber compounds, it will be necessary to watch for leaks during the first few weeks of operating the system. Retighten the clamps as necessary.

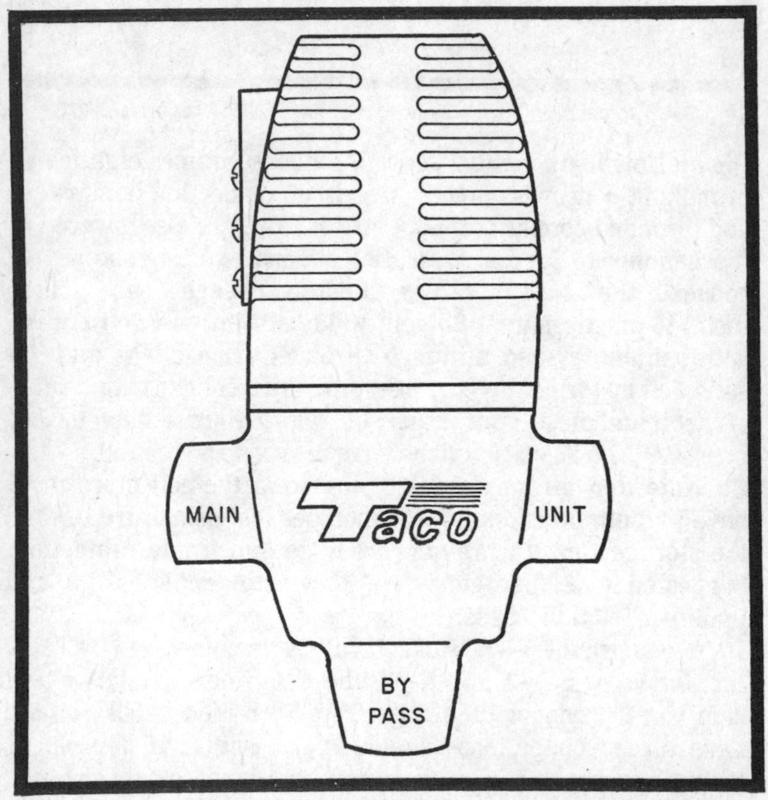

Fig. 9-5. Location of connections for Taco valve models 561 and 562.

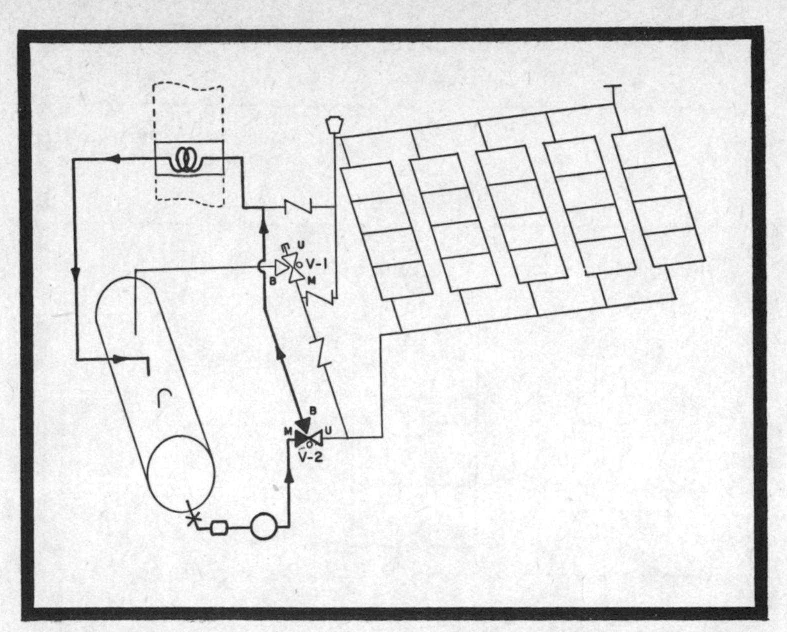

Fig. 9-6. Schematic for connecting Taco valves in typical plumbing layout. Water flow path shown when valves are de-energized.

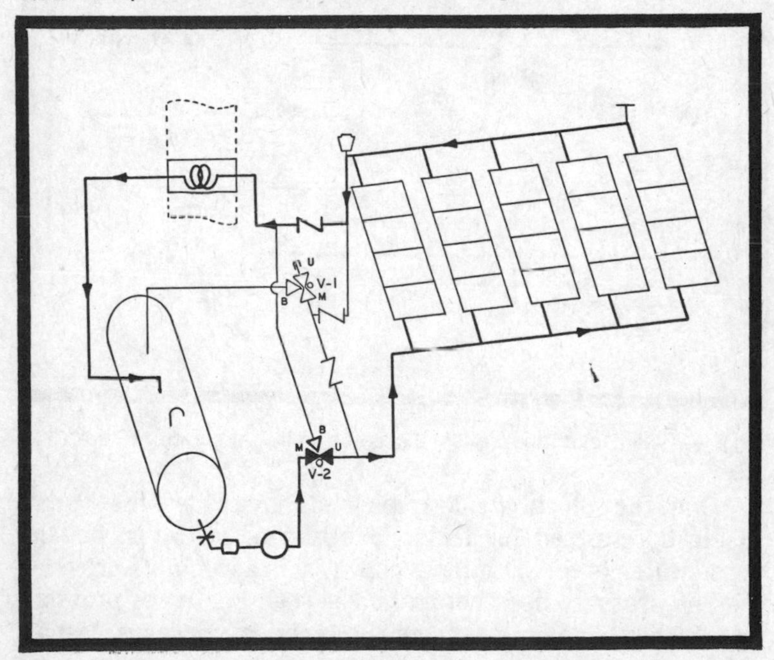

Fig. 9-7. Water flow path through collectors with valves in energized position.

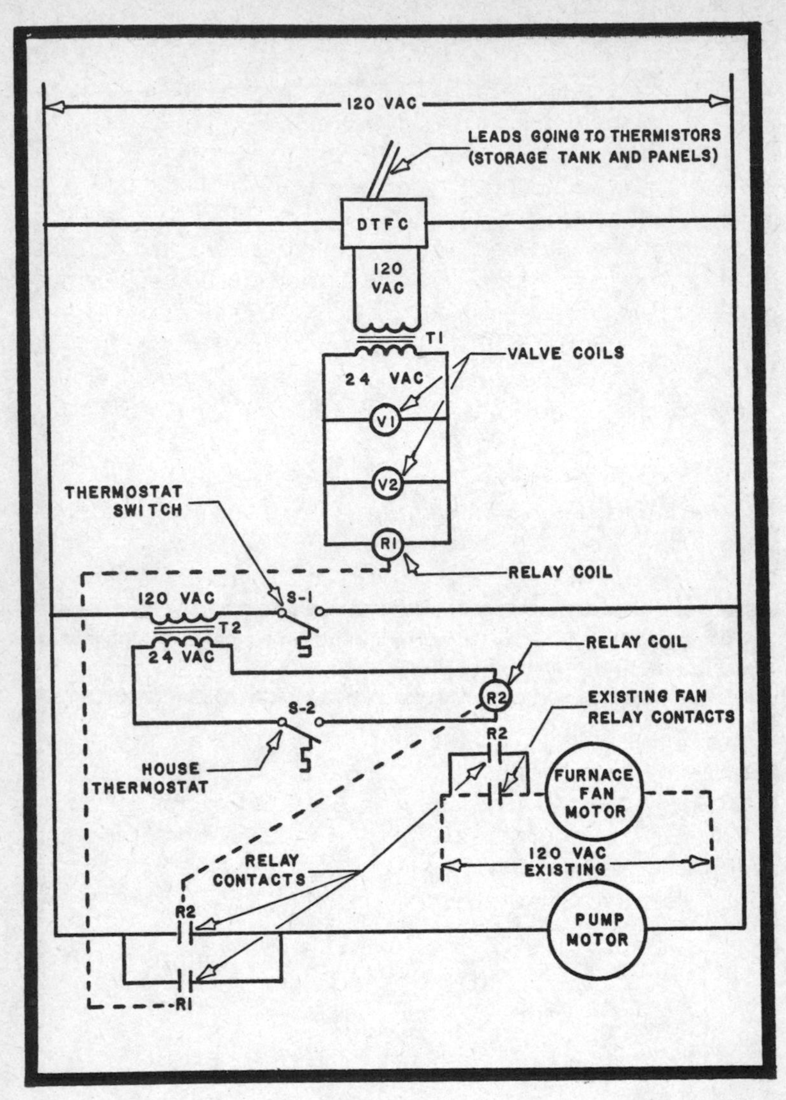

Fig. 9-8. Typical electrical schematic as shown in NASA report N76-27671.

Once the solvent-welded pipe joints have cured, the system should be checked for leaks. One method is to connect the house water to the plumbing loop. This is acceptable provided the line pressure does not exceed 100 psi. Leave the pressure on for 8 hours and make several inspections for leaks. Repair it if necessary after draining, using new fittings and pipe splices as required.

ELECTRICAL CONNECTIONS

Electrical components are shown in schematic form in Fig. 9-8. You may wish to refer to Fig. 7-2 for a description of the symbols used, and to the bill of materials provided in this chapter for component descriptions keyed to this schematic.

It is recommended that you consider an electrical contractor for this work. If you understand the circuits and have the necessary skills, you may be able to obtain a permit, do the work, and satisfy the local electrical inspector. In either case, NASA manual N76-27671, mentioned previously, is most helpful in providing a more thorough explanation of the electrical circuit functions.

WATER TREATMENT

Untreated tap water can cause premature corrosion in aluminum absorbers. Water chemistry varies widely throughout the United States, so the best procedure is to contact a local water treatment company when you are ready to fill the system. The company will recommend corrosion inhibitors and additives necessary to maintain a proper PH balance. Be sure to list the various metals used in the system that the water will contact, such as aluminum, copper, and steel. The water chemistry should be periodically checked as makeup water is added to the tank to compensate for evaporation, minor leakage, etc. The treatment company can suggest a schedule. If you ignore water treatment, there is a good possibility your system will fail prematurely.

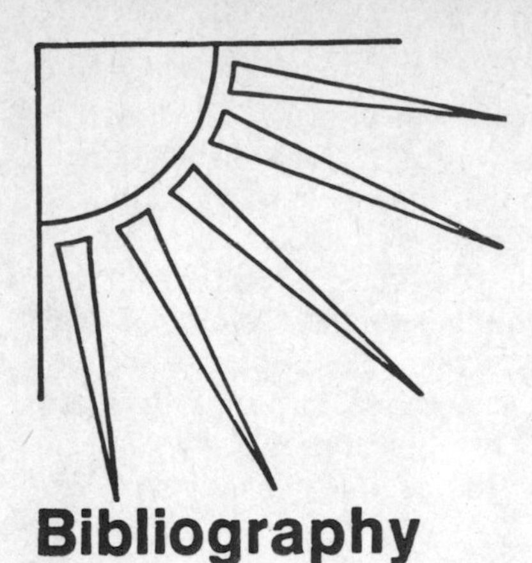

Bibliography

For those of you who are interested in securing more information about solar energy, the following sources are provided. This is by no means a complete source of information, as more data are being released and more research is being done every day. But these are the sources I found useful in preparing this book, and I am sure they will be helpful to you.

Allred, Johnny W.; Barringer, Sheridan R.; Kirby, Cecil E.; Shinn, Joseph M., Jr. *An Inexpensive Economical Solar Heating System for Homes.* NASA Langley Research Center, Hampton, Virginia. Prepared for National Technical Information Service, Springfield, Virginia, N76-27671. July 1976.

Balcomb, J. Douglas; Hedstrom, James C. *A Simplified Method for Sizing a Solar Collector Array for Space Heating.* Los Alamos Scientific Laboratory of the University of California, Los Alamos, New Mexico. Prepared for U.S. Energy Research and Development Administration, LA-UR-76-160. January 23, 1976.

Borzoni, J.T.; Holland, T.H.; Ramsey, J.W. *Development of Flat-Plate Solar Collectors for the Heating and Cooling of Buildings.* Honeywell Inc., Systems and Research Center, Minneapolis, Minnesota. Prepared for the National Aeronautics and Space Administration, NASA CR-134804. June 1975.

Climatic Atlas of the United States. U.S. Department of Commerce. 1974.

Cook, Jerry E., managing director. Grundfos Pumps Corporation, Clovis, California. Personal communication, January 20, 1977.

Davis, Andrew D., vice president. Rho Sigma Inc., Van Nuys, California. Personal communication, January 17, 1977.

DeSoto Inc., Des Plaines, Illinois. Enersorb data.

Development of Proposed Standards for Testing Solar Collectors and Thermal Storage Devices. National Bureau of Standards, Technical Note 899. Prepared for the Energy Research and Development Administration, Division of Solar Energy, Washington D.C. February 1976.

de Winter, Francis. *How to Design and Build a Solar Swimming Pool Heater,* Technical Report. Copper Development Association Inc., New York, New York. 1975.

Drury, William B., product manager. Taco Inc., Cranston, Rhode Island. Personal communication, March 14, 1977 and March 22, 1977.

General Electric Co., Silicone Products Department, Waterford, New York. Silglaze Silicone Glazing System data. May 1975.

Glazing Manual. Flat Glass Marketing Association, Topeka, Kansas. 1974.

Interim Performance Criteria for Solar Heating and Combined Heating/Cooling Systems and Dwellings. National Bureau of Standards, U.S. Department of Commerce, Washington, D.C. Prepared for U.S. Department of Housing and Urban Development. January 1, 1975.

Intermediate Minimum Property Standards for Solar Heating and Domestic Hot Water Systems. Solar Energy Program Team, National Bureau of Standards, Center for Building Technology. Prepared for U.S. Department of Housing and Urban Development, Division of Energy, Building Technology and Standards, Washington, D.C., NBSIR 76-1059. April 1976.

Johns-Manville Sales Corporation, Denver, Colorado. Spin-Glas data.

Lior, N.; Saunders, A.P. *Solar Collector Performance Studies*. University of Pennsylvania. National Center for Energy Management and Power, Philadelphia, Pennsylvania. Prepared for the National Science Foundation, NSF-RA-N-74-152(3). 1974.

Low-Temperature Engineering Application of Solar Energy. ASHRAE members of technical committee on solar energy utilization. HS-5-75-1000. 1967.

Melicher, Ronald W.; Sciglimpaglia, Donald M.; Scott, Jerome E. *Demand Analysis, Solar Heating and Cooling of Buildings—Solar Water Heating in South Florida: 1923-1974*. Phase I Report, prepared for the National Science Foundation, NSF-RA-N-74-190. December 1974.

Mumpton, Kurt L., product manager. Taco Inc., Cranston, Rhode Island. Personal communication, February 23, 1977.

NIBCO Copper Piping Manual, CPM-2; *The NIBCO Plastic Piping How-to-and-why Pocket Handbook*. NIBCO Inc., Elkhart, Indiana. 1976.

Solar Energy Engineering and Product Catalog. Rho Sigma Inc., Van Nuys, California.

Solar Energy Utilization for Heating and Cooling. ASHRAE Handbook and Product Directory, Applications Volume, Chapter 59. The American Society of Heating, Refrigerating, and Air Conditioning Engineers Inc. 1974.

Solar Engineering Magazine. Official publication of the Solar Energy Industries Association, monthly. Dallas, Texas.

Solar Heating and Cooling Experiment for a School in Atlanta. Westinghouse Electric Corporation, Special Systems, Baltimore, Maryland; Burt, Hill & Associates, Architects, Butler, Pennsylvania. NSF-C-908. December 1, 1974.

Solar Heating and Cooling of Buildings. TRW Systems Group, Redondo Beach, California. Phase O, Volume II, Final Report. Prepared for the National Science Foundation, NSF-RA-N-74-022B. May 31, 1974.

Solar Heating Handbook for Los Alamos. Los Alamos Scientific Laboratory of the University of California, Los Alamos, New Mexico. LA-5967. May 1975.

Sosnin, H.A. *The Theory and Technique of Soldering and Brazing of Piping Systems*. Illustrations by Walter K. Hood, NIBCO Inc. 1971.

Stern, Jack *A 5-Minute Insulating Glass Seminar*. Hygrade Metal Moulding Manufacturing Corp., Farmingdale, New York.

Szokolay. S.V. *Solar Energy and Building*. John Wiley & Sons. New York, New York. 1975.

Taco Inc., Cranston, Rhode Island. SSM Solar Systemizer data. Capalog 100-4.

Uniform Solar Energy Code. International Association of Plumbing and Mechanical Officials. Los Angeles, California. September 1976.

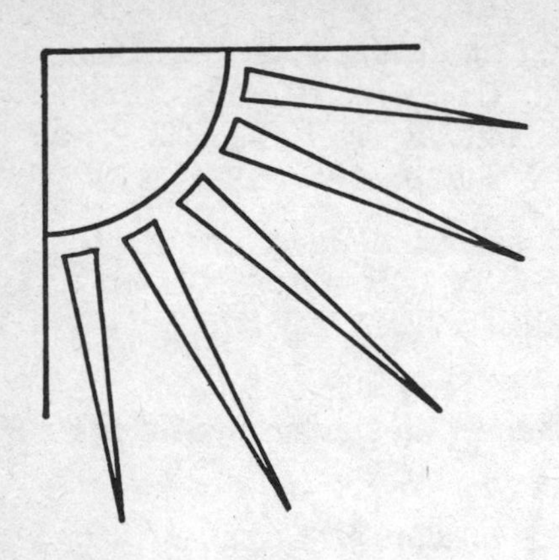

Index